# FUN WITH NUMBERS

Amit Garg

PUSTAK MAHAL®

*Publishers*
**Pustak Mahal**®

J-3/16 , Daryaganj, New Delhi-110002
☎ 23276539, 23272783, 23272784 • *Fax:* 011-23260518
*E-mail:* info@pustakmahal.com • *Website:* www.pustakmahal.com

***Sales Centre***

- 10-B, Netaji Subhash Marg, Daryaganj, New Delhi-110002
  ☎ 23268292, 23268293, 23279900 • *Fax:* 011-23280567
  *E-mail:* rapidexdelhi@indiatimes.com
- **Hind Pustak Bhawan**
  6686, Khari Baoli, Delhi-110006
  ☎ 23944314, 23911979

***Branches***

**Bengaluru:** ☎ 080-22234025 • *Telefax:* 080-22240209
*E-mail*: pustak@airtelmail.in • pustak@sancharnet.in
**Mumbai:** ☎ 022-22010941, 022-22053387
*E-mail*: rapidex@bom5.vsnl.net.in
**Patna:** ☎ 0612-3294193 • *Telefax:* 0612-2302719
*E-mail*: rapidexptn@rediffmail.com
**Hyderabad:** *Telefax:* 040-24737290
*E-mail*: pustakmahalhyd@yahoo.co.in



ISBN 978-81-223-0352-0

**Edition : 2012**



***Printed at :*** Param Offsetters, Okhla, Delhi-110020

# CONTENTS

R. GHOSH

1

# NUMBER RECREATIONS

**1. The sum of a square of three numbers in a calendar**

Ask your friend to choose three successive dates from a calendar. Let him do so in a column starting with the lowest number. You can tell the sum of 9 digits of the square so formed (Fig. 1.1). How? Ask your friend the least number he had thought of. Add 8 to it and multiply by 9. This will yield the desired sum.

| S | M | T | W | T | F | S |
|---|---|---|---|---|---|---|
| | | | | 1 | 2 | 3 |
| 4 | 5 | 6 | 7 | 8 | 9 | 10 |
| 11 | 12 | 13 | 14 | 15 | 16 | 17 |
| 18 | 19 | 20 | 21 | 22 | 23 | 24 |
| 25 | 26 | 27 | 28 | 29 | 30 | 31 |

*Fig. 1.1*

**2. The age detector table**

The table on the following page helps you in guessing the age. What you have to do is to ask your friend about the columns where his age appears. Simply add up the numbers at the top of the columns indicated, and that sum will be equal to the age of your friend.

For example, your friend's age is 23. This number as you can see is available in first, second, third and fifth columns. The top numbers in these four columns are 1, 2, 4, 16—which add up to 23.

Table

| 1 | 2 | 4 | 8 | 16 | 32 |
|---|---|---|---|---|---|
| 3 | 3 | 5 | 9 | 17 | 33 |
| 5 | 6 | 6 | 10 | 18 | 34 |
| 7 | 7 | 7 | 11 | 19 | 35 |
| 9 | 10 | 12 | 12 | 20 | 36 |
| 11 | 11 | 13 | 13 | 21 | 37 |
| 13 | 14 | 14 | 14 | 22 | 38 |
| 15 | 15 | 15 | 15 | 23 | 39 |
| 17 | 18 | 20 | 24 | 24 | 40 |
| 19 | 19 | 21 | 25 | 25 | 41 |
| 21 | 22 | 22 | 26 | 26 | 42 |
| 23 | 23 | 23 | 27 | 27 | 43 |
| 25 | 26 | 28 | 28 | 28 | 44 |
| 27 | 27 | 29 | 29 | 29 | 45 |
| 29 | 30 | 30 | 30 | 30 | 46 |
| 31 | 31 | 31 | 31 | 31 | 47 |
| 33 | 34 | 36 | 40 | 48 | 48 |
| 35 | 35 | 37 | 41 | 49 | 49 |
| 37 | 38 | 38 | 42 | 50 | 50 |
| 39 | 39 | 39 | 43 | 51 | 51 |
| 41 | 42 | 44 | 44 | 52 | 52 |
| 43 | 43 | 45 | 45 | 53 | 53 |
| 45 | 46 | 46 | 46 | 54 | 54 |
| 47 | 47 | 47 | 47 | 55 | 55 |
| 49 | 50 | 52 | 56 | 56 | 56 |
| 51 | 51 | 53 | 57 | 57 | 57 |
| 53 | 54 | 54 | 58 | 58 | 58 |
| 55 | 55 | 55 | 59 | 59 | 59 |
| 57 | 58 | 60 | 60 | 60 | 60 |
| 59 | 59 | 61 | 61 | 61 | 61 |
| 61 | 62 | 62 | 62 | 62 | 62 |
| 63 | 63 | 63 | 63 | 63 | 63 |

3. **Number of brothers and sisters one had — if both the parents were alive**

Here is a simple trick by which you can tell the number of brothers and sisters one had.

*Step 1* — Add the number of parents (obviously it will be two in each case).

*Step 2* — Add to Step 1 the number of brothers and multiply the result by 2 and add one to it.

*Step 3* — Now the result of Step 2 is multiplied by 5 and the number of sisters is added to it.

*Step 4* — Subtract 25 from Step 3 to get the result.

*Step 5* — In the result of Step 4 the unit's place indicates the number of sisters and the ten's place the number of brothers.

Hence we can write the formula as :

$$[(\text{parents} + \text{brothers}) \times 2 + 1] \times 5 + \text{sisters} - 25$$

**Example:**

Suppose Ram has 3 brothers and two sisters, then the solution would be:

| | | | |
|---|---|---|---|
| *Step 1* — | Number of parents | = | 2 |
| *Step 2* — | $(2 + 3) \times 2 + 1$ | = | 11 |
| *Step 3* — | $11 \times 5 + 2$ | = | 57 |
| *Step 4* — | $57 - 25$ | = | 32 |
| *Step 5* — | Unit's place | = | 2 sisters |
| | Ten's place | = | 3 brothers |

4. **The age from the phone number**

To tell the age of anyone, ask him to write down last four digits of a telephone number. Suppose the four digits are 8231. Let him conceive a new four-digit number by placing the above digits a new order, say 2318. Now, subtract the smaller number from the bigger one. In this case the result is 5913.

Add all the digits i.e. $5 + 9 + 1 + 3 = 18 = 1 + 8 = 9$

Add sixteen to the result i.e. $9 + 16 = 25$

Now ask the person to add this number to his birth year, i.e. 1945 + 25 = 1970

Now ask him about the final result. You can tell his birth year by subtracting 25 from the final result i.e. 1970 – 25 = 1945.

**5. The age of anybody**

If someone is acquainted with the basics of arithmetic, he can tell the age of anybody by following the simple method given below:

***Step 1*** — Tell him/her to add his/her present age and the next year's age.

***Step 2*** — Tell him/her to multiply the sum by 5.

***Step 3*** — Tell him/her to add the unit digit of the year in which he/she was born.

***Step 4*** — Tell him/her to subtract 5 from the final result.

***Step 5*** — The left two digits will indicate his/her age.

**Example**

Suppose today your friend is 12 years old and he was born in 1976.

***Step 1*** — Present age + next year's age = 25

***Step 2*** — Step 1 × 5 i.e. 25 × 5 = 125

***Step 3*** — Step 2 + unit digit of the year in which your friend was born i.e. 125 + 6 = 131

***Step 4*** — Step 3 – 5 = 131 – 5 = 126

***Step 5*** — The left two digits are 12, hence his age.

**6. The number of currency notes in one's pocket.**

1. Ask him to write down the number of currency notes in his pocket.
2. Then double it.
3. Add 1 to the above number.
4. Multiply the above number by 5.
5. Now add 5 to the above result.

6. Next ask him to multiply the result by 10.
7. Subtract 100 from the above number.
8. Now ask him to tell you the number. Delete the last two digits from this number and you will get the number of currency notes eventually.

**Example**

Suppose your friend writes down the number of currency notes as: : 16

On doubling it becomes : 32

Then he adds 1 to the above number : 33

He multiplies the above number by 5 : 165

He adds 5 to the above result : 170

Next he multiplies the number by 10 to get : 1700

He then subtracts 100 from the above number to get : 1600

Now ask him to tell you the number after deleting the last two digits from the result (i.e. 1600), you can tell that he has 16 currency notes in his pocket.

**7. A wonder with the date of birth and age**

Let the days of the week and the months of the year be represented by numbers. For the days of the week,

Let 1. represents Monday

2. represents Tuesday

3. represents Wednesday, and so on.

Similarly, for the months of the year,

Let 1. represents January

2. represents February

3. represents March and so on.

1. Let your friend write down the number that represents the day of the week, month of the year and the date of the month in which he was born.
2. Ask him to double this number.

3. Ask him to add 5 to the above number.
4. Ask him to multiply the addendum by 50.
5. Add his present age to the above result.
6. Ask him to subtract 365 from the above.
7. Let him add 115 to the above.

The result is miraculous, since the answer gives first — the day of the week on which your friend was born, next the month in which he was born, next the date of the month and finally his present age.

**Example**

Suppose your friend was born on Sunday 6th May 1962 and is now 23 years old.

1. We start with the number 756 where 7 represents the day of the week, 5 the month of the year and 6 the date of the month.
2. Ask your friend to double the number : 1512.
3. Let him add 5 to the above number : 1517.
4. Let him multiply the above number by 50 : 75850.
5. Let him add his present age : 75873.
6. Let him subtract 365 from the above number : 75508.
7. Eventually, let him add 115 to arrive at : 75623.

Here, the first number gives the day of the week on which he was born i.e. Sunday. The second number gives month of the year i.e May. Third number gives the date of the month i.e., 6th. Finally, we shall arrive at his present age as 23.

**8. Tell your pal's age and the coins (less than a rupee) he had**

Insist your pal to proceed in this way.

1. Let him conceive his age in mind.
2. Then say to double it.
3. Add 5.
4. Multiply by 50.

5. Subtract 365.
6. Add to this the change in the pocket.
7. Add 115.

Whatever result he arrives at, the first two figures will be indicating his age and the last two the change he had in his pocket.

**For example**

| | | | |
|---|---|---|---|
| 1. | Say friend's age | = | 20 |
| 2. | Double | = | 40 |
| 3. | Add 5 | = | 45 |
| 4. | Multiply by 50 | = | 2,250 |
| 5. | Subtract 365 | = | 1,885 |
| 6. | Add change (let it be) 75 paise | = | 1,960 |
| 7. | Add 115 | = | 2,075 |

2,075 indicates his age as 20 years and number of coins as 75 paise.

**9. Number of brothers and sisters of someone whose parents are hardly alive.**

There are many persons in the world, whose parents are not alive. By the following trick you can foretell the number of brothers and sisters they had.

*Step 1* — Multiply the number of brothers by 2 and add 3 to it.

*Step 2* — Multiply the result of Step 1 by 5 and later add the number of sisters.

*Step 3* — Finally, subtract 15 from the result of Step 2.

*Step 4* — In the final result, the unit's place digit indicates the number of sisters while the ten's place digit indicates the number of brothers.

This trick has one limitation — if the number of brothers and sisters is more than 9, it can't be applied. The method is valid for the followers of family planning. ■■

# 2

# NUMBER TRICKS

1. **Getting a three-digit integer written twice**

   Consider any three-digit integer. Multiply it by 11, and the overall answer by 91. Now, the result is the original integer written twice.

   **Example 1**

   1. Consider the three-digit integer, say : 456
   2. Multiplying this by 11 : 5016
   3. Multiplying the multiplicand by 91 : 456 456

      which is the original integer written twice in apposition (written side by side)

   **Example 2**

   1. Consider any other three-digit integer, say : 777
   2. Multiplying this by 11 : 8547
   3. Multiplying the above number by 91 gives : 777777

      which again is the original integer written twice in apposition (written side by side).

2. **Jugglery of three-digit numbers**

   1. Consider any three-digit number.
   2. Repeat this three-digit number to form a six-digit one in apposition.
   3. Divide the overall number by 7.
   4. Divide the dividend by 11.
   5. Divide the factorial by 13.
   6. You will arrive at the same number you started.

**Example 1**

1. Consider the three-digit number : 362
2. Now repeat this three-digit number to form a six-digit number : 362, 362
3. Dividing it by 7, we get : 51,766
4. Dividing 51,766 by 11 gives : 4701
5. Dividing 4701 by 13 eventually : 362

**Example 2**

1. Consider the three-digit number : 789
2. Now repeat this three-digit number to form a six-digit number : 789, 789
3. Dividing it by 7, we get : 112,827
4. Dividing 112827 by 11 gives : 10,257
5. Dividing 10257 by 13 eventually : 789

**3. Back to the same number 7**

Pause over any number. Now double that. Add 5 to it. Add 12 more to it and subtract 3 and divide by 2. Now subtract the original number from the above result. The result is always 7.

**Example 1**

1. Let us consider the number : 10
2. Doubling it, we get : 20
3. Now adding 5 and 12 successively, we get : 25 & 37
4. Subtracting 3 from the above yields : 34
5. Dividing the above number by 2 would make : 17
6. Subtracting 10 (which was the original number) from the above gives : 7

**Example 2**

1. Let us consider any other number, say : 45
2. Doubling this number, we get : 90
3. Adding 5 to the above number and 12 successively, we get : 95 & 107

4. Subtracting 3 from above number gives : 104
5. Dividing the above number by 2 gives : 52
6. Subtracting 45 from the above gives : 7

The answer is always 7. You can try the trick.

**4. An amazing number trick**

Consider any number. Multiply this number by 3, and add 2 to the above result. Multiply the overall by 3. Add a number that is two more than the number initially thought of. The number after the unit digit in the final answer will always be the number initially conceived.

**Example 1**

1. Let us consider the number as : 35
2. Multiplying this number by 3, it gives : 105
3. Adding 2 to the above number, we get : 107
4. Multiplying again by 3, we get : 321
5. Now adding a number which is 2 more than the number first thought of (since 35 was the number initially thought of, we must add 37) : 358

Now the number after the unit digit in the answer i.e. 35 remains the number initially meant.

**Example 2**

1. Let us consider any other number, say : 7
2. Multiplying this number by 3 gives : 21
3. Adding 2 to the above number we get : 23
4. Multiplying the above number by 3 we get : 69
5. Now adding 9 to the above number (since, it is two more than the original number) : 78

You will find that the number after the unit digit in the answer is the number initially thought of i.e., 7.

**5. Magic number**

Here is another number which will give you a lot of surprises.

This magic number is 142857. Look at these —

$142857 \times 2 = 285714$

$142857 \times 3 = 428571$

$142857 \times 4 = 571428$

$142857 \times 5 = 714285$

$142857 \times 6 = 857142$

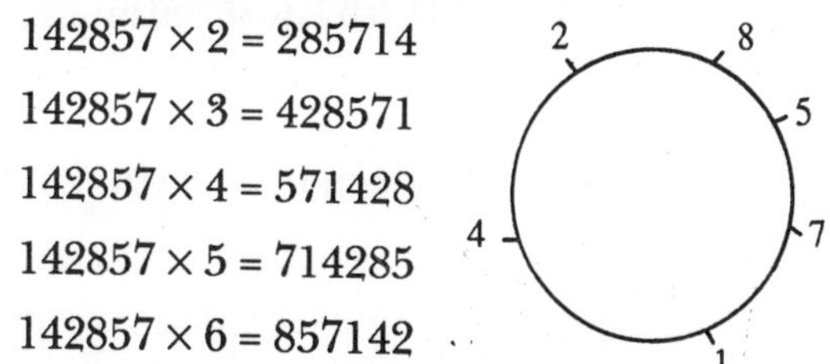


It will be very clear to you by now that if you multiply 142857 by 2, 3, 4, 5, 6 you will get same figures in the same order, starting in a different place each time as if they were written round the edge of a circle.

**6. Tell the missing digit**

***Step 1*** — Choose a large number of six or seven digits.

***Step 2*** — Take the sum of digits.

***Step 3*** — Subtract sum of digits from any number chosen.

***Step 4*** — Mix up the digits of resulting number.

***Step 5*** — Add 25 to it.

***Step 6*** — Cross out any one digit except zero.

***Step 7*** — Tell the sum of the digits.

From the sum of digits you can tell the digit crossed out.

In order to find out the missing digit, subtract the sum of digits from 25. The difference is the missing digit.

**Example**

| | | |
|---|---|---|
| ***Step 1*** — Number chosen | : | 1234567 |
| ***Step 2*** — Sum of digits | : | 28 |
| ***Step 3*** — 1234567 – 28 | : | 1234539 |
| ***Step 4*** — | : | 1235349 |
| ***Step 5*** — 1235349 + 25 | : | 1235374 |
| ***Step 6*** — Cross out digit | : | 5 |
| ***Step 7*** — The sum of digits | : | 20 |
| Number crossed out | : | 25 – 20 = 5 |

**7. The predicted total of 1089**

Here is an amusing trick with numbers. Tell your friend that you can predict his answer on a piece of paper. You have to write 1089 on a piece of paper. Let him put it in his pocket where you hardly touch it. The game begins now.

Let him write any three-digit number. It should be such that the left hand or hundred's digit exceeds the right hand or unit's digit by two or beyond.

Suppose he assumes 981. Tell him to write it down backwards underneath his first number. That would make it 189. Tell him to subtract the smaller one from the larger. It will be 792.

He had tc add to the result the same number reversed i.e. 792 + 297. The result is 1089. The process moves as:

| | | |
|---|---|---|
| Number assumed | 761 | 321 |
| After reversal | 167 | 123 |
| Subtraction (smaller one from larger) | 594 | 198 |
| Answer reversed | 495 | 891 |
| Addition | 1089 | 1089 |

When your friend opens your predicted tag and sees that you have written 1089 he shall wonder how you arrived at, ahead of time.

**8. To tell an integer from one digit**

Think a number between 1 and 9.

Annex a zero to it.

Add the following number :

Multiply this sum by 3, 11 and 3 successively.

Ask the last digit of the answer.

From this you can tell complete number as follows:

— You know the last digit.

— For second digit, subtract the last digit from 9.

— The first digit will be one more than the second digit. This is the first digit of the original number.

— The third digit will be the first digit subtracted from 9.

Arrange this and the complete number is with you.

**Example**

| | |
|---|---|
| Number thought | = 3 |
| Annexing zero | = 30 |
| Adding the number | = 33 |
| Multiplying by 3 | = 99 |
| Multiplying by 11 | = 1089 |
| Multiplying by 3 | = 3267 |
| — Last digit | 7 |
| — Second digit | (9 – 7) = 2 |
| — First digit | (2 + 1) = 3 |
| — Third digit | (9 – 3) = 6 |

9. **Secret digit**

One can tell the secrecy of digit 9 (integer) in the following way:

*Step 1* — Let your friend think of a six-digit number.

*Step 2* — Let him add the digits.

*Step 3* — Let him subtract sum of digits from that of six digit number.

*Step 4* — Tell him that he can keep one number secret and could encircle it.

*Step 5* — Tell him to add the remaining digits barring that secret number.

*Step 6* — When he reveals the result, you can tell him the secret digit by subtracting that number from the next product of 9.

**Example 1**

| | | |
|---|---|---|
| *Step 1* — | Six digit number | 124567 |
| *Step 2* — | Sum of digits | 1 + 2 + 4 + 5 + 6 + 7 = 25 |

*Step 3* — Six-digit No – sum of digits 124567 – 25 = 124542

*Step 4* — Let the secret number = 5

*Step 5* — Sum of the remaining digits = 13

*Step 6* — So the final result is 13, and you just subtract 13 from next product of 9 i.e. 18, you will end up with 5 which is the secret digit of your friend.

**Example 2**

*Step 1* — Six-digit number 549218

*Step 2* — Sum of digits 29

*Step 3* — Six-digit number – sum of digits 549189

*Step 4* — Let the secret digit be 1

*Step 5* — Sum of remaining digits 35

*Step 6* — Subtract 35 from the next product of 9 which will be 36 and hence 36 – 35 = 1, which yields the secret digit of your pal.

**10. Integral number guesser**

*Step 1* — Tell your crony to think two numbers between 1 and 9.

*Step 2* — Double the first number and add 5 to it.

*Step 3* — Multiply the result of Step 2 by 5 and tell him to add the second number.

*Step 4* — Result of Step 3–25.

*Step 5* — The result of Step 4 will give you the numbers thought by your crony.

**Example**

*Step 1* — Let the number thought be 1 and 8.

*Step 2* Double value of 1st number + 5 = 2 × 1 + 5 = 7.

*Step 3* — Result of step 2×5 + second No=7×5 +8 =35 +8 =43.

*Step 4* — Result of Step 3 – 25 = 43 – 25 = 18.

*Step 5* — Result of Step 4=18 which shows 1 and 8 are the numbers thought by your crony.

**Another method**

*Step 1* — Tell your crony to think of an integer, say a two-digit number.

*Step 2* — Tell him to multiply the ten's place digit by 5 and then add 7 in it.

*Step 3* — Double the result of Step 2.

*Step 4* — Add the unit's digit in the result of Step 3 and ask your friend the result.

*Step 5* — Subtract silently 14 from the result without telling him. Following the subtraction, the remaining one will be a two digit number thought of by your crony.

**Example**

*Step 1* — Let the number be 27.

*Step 2* — $2 \times 5 + 7 = 17$.

*Step 3* — Double of Step 2 = $2 \times 17 = 34$.

*Step 4* — Adding unit's digit to Step 3 result = $7 + 34 = 41$.

*Step 5* — Subtracting 14 will give $41 - 14 = 27$ which is the number thought of by your crony.

**Parallel method**

*Step 1* — Tell your crony to think an integer, say a three-digit number.

*Step 2* — Multiply hundred's place digit by 2 and add three to it.

*Step 3* — Add 7 to the multiplicand got out by multiplier 5.

*Step 4* — Add the tenth place digit of the original number to the result of Step 3.

*Step 5* — Double the result of Step 4 and add 3.

*Step 6* — Multiply the result of Step 5 by 5, add the unit place digit of original number to it.

*Step 7* — Ask the final result from your crony and in your mind subtract 235 from it. The result will be the number thought of.

**Example**

*Step 1* — Let the number thought of be = 100

*Step 2* — $1 \times 2 + 3$ = 5

*Step 3* — $5 \times 5 + 7$ = 32

*Step 4* — $32 + 0$ = 32

*Step 5* — $32 \times 2 + 3$ = 67

*Step 6* — $67 \times 5 + 0$ = 335

*Step 7* — $335 - 235 = 100$ which is the number thought of by your crony.

**11. A number game, where the answer hangs upon 15**

Pause upon some number say 12, multiply it by 5, making 60. Add 25 to it, making 85. Divide it by 5, making the quotient 17. Subtract the number you started with $17 - 12 = 5$. Multiply the remainder by 3 making it 15. So the answer will remain as 15.

**12. Time to count a billion**

If a man counts at the rate of 100 numbers a minute, and kept doing for eight hours a day, five days a week, it would take a little over four weeks to count a million and just over 80 years to reach a billion. That's life indeed, practically. ■■

# 3

# NUMBER PATTERNS

Natural numbers have made some fascination in our practical mundane life. If one delves with some curiosity, he shall find only their structure bit objective, unable to trod on with their natural phenomena. Mathematicians are finding themselves at odds, despite their futility and loss of patience, though they have advanced to the age of nuclear war in three dimensional space.

The pattern is neither arithmetic nor geometric nor harmonic in progression but recurs up to infinity (ad infinitum). Mathematicians were quite goofy to frame it, but hardly had they attributed any concept. Natural numbers have their own structure, by virtue. Let's jog on to natural number 8.

**1. Trick with natural number 8**

$1 \times 8 + 1 = 9$

$12 \times 8 + 2 = 98$

$123 \times 8 + 3 = 987$

$1234 \times 8 + 4 = 9876$

$12345 \times 8 + 5 = 98765$

$123456 \times 8 + 6 = 987654$

$1234567 \times 8 + 7 = 9876543$

$12345678 \times 8 + 8 = 98765432$

$123456789 \times 8 + 9 = 987654321$

**2. Trick with natural number 9 (Pattern of 11)**

$0 \times 9 + 1 = 1$

$1 \times 9 + 2 = 11$

$12 \times 9 + 3 = 111$

$123 \times 9 + 4 = 1111$

$1234 \times 9 + 5 = 11111$

$12345 \times 9 + 6 = 111111$

$123456 \times 9 + 7 = 1111111$

$1234567 \times 9 + 8 = 11111111$

$12345678 \times 9 + 9 = 111111111$

$123456789 \times 9 + 10 = 1111111111$

**3. Fun with multiplication**

Multiply the number below and you will get the magical results:

$9 \times 0 + 8 = 8$

$9 \times 9 + 7 = 88$

$9 \times 98 + 6 = 888$

$9 \times 987 + 5 = 8888$

$9 \times 9876 + 4 = 88888$

$9 \times 98765 + 3 = 888888$

$9 \times 987654 + 2 = 8888888$

$9 \times 9876543 + 1 = 88888888$

$9 \times 98765432 + 0 = 888888888$

So by this you have got a lot of 8's.

**4. Fun with 37**

If one multiplies 37 by 3, 6, 9, 12, 15, 18, 21, 24, 27 (Arithmetic progression) then the wonderful results given by it are:

$37 \times 3 = 111$

$37 \times 6 = 222$

$37 \times 9 = 333$

$37 \times 12 = 444$

$37 \times 15 = 555$

$37 \times 18 = 666$

$37 \times 21 = 777$

$37 \times 24 = 888$

$37 \times 27 = 999$

**5. Fascinating results of 9**

| | |
|---|---|
| $9 \times 9$ | $= 81$ |
| $99 \times 99$ | $= 9801$ |
| $999 \times 999$ | $= 998001$ |
| $9999 \times 9999$ | $= 99980001$ |
| $99999 \times 99999$ | $= 9999800001$ |

**6. Another trick with number 9**

| | |
|---|---|
| $987654321 \times 9 - 1$ | $= 8888888888$ |
| $87654321 \times 9 - 1$ | $= 788888888$ |
| $7654321 \times 9 - 1$ | $= 68888888$ |
| $654321 \times 9 - 1$ | $= 5888888$ |
| $54321 \times 9 - 1$ | $= 488888$ |
| $4321 \times 9 - 1$ | $= 38888$ |
| $321 \times 9 - 1$ | $= 2888$ |
| $21 \times 9 - 1$ | $= 188$ |
| $1 \times 9 - 1$ | $= 8$ |

**7. Trick with number 7**

| | |
|---|---|
| $7 \times 7$ | $= 49$ |
| $67 \times 67$ | $= 4489$ |
| $667 \times 667$ | $= 444889$ |
| $6667 \times 6667$ | $= 44448889$ |
| $66667 \times 66667$ | $= 4444488889$ |
| $666667 \times 666667$ | $= 444444888889$ |

**8. Trick with 3 and 4**

| | |
|---|---|
| $4 \times 4$ | $= 16$ |
| $34 \times 34$ | $= 1156$ |
| $334 \times 334$ | $= 111556$ |
| $3334 \times 3334$ | $= 11115556$ |
| $33334 \times 33334$ | $= 1111155556$ |
| $333334 \times 333334$ | $= 111111555556$ |

**9. Amusing trick**

123456789 × 9 = 111111101

123456789 × 18 = 222222202

123456789 × 27 = 333333303

123456789 × 36 = 444444404

123456789 × 45 = 555555505

123456789 × 54 = 666666606

123456789 × 63 = 777777707

123456789 × 72 = 888888808

123456789 × 81 = 999999909

**10. Strange multiplication**

987654321 × 9 = 8888888889

987654321 × 18 = 17777777778

987654321 × 27 = 26666666667

987654321 × 36 = 35555555556

987654321 × 45 = 44444444445

987654321 × 54 = 53333333334

987654321 × 63 = 62222222223

987654321 × 72 = 71111111112

987654321 × 81 = 80000000001

**11. Magic of nine**

222222222 × 9 = 1999999998

333333333 × 9 = 2999999997

444444444 × 9 = 3999999996

555555555 × 9 = 4999999995

666666666 × 9 = 5999999994

777777777 × 9 = 6999999993

888888888 × 9 = 7999999992

999999999 × 9 = 8999999991

**12. Super powers**

Squaring each of the natural number in the logarithmic fashion, one will get recurrence of natural numbers in tandem.

$1^2 = 1$

$11^2 = 121$

$111^2 = 12321$

$1111^2 = 1234321$

$11111^2 = 123454321$

$111111^2 = 12345654321$

$1111111^2 = 1234567654321$

$11111111^2 = 123456787654321$

$111111111^2 = 12345678987654321$

**13. Get a series of '2'**

If you want to get series of 2 just by adding numbers do the following.

Write natural numbers from 1 to 9 at a stretch and then 9 to 1 in descending order and again 1 to 9 on the third row in ascending order and 9 to 1 in the fourth in descending order.

In the fifth just write 2 and add. Then you get a series of 2 in tandem.

| | | | | | | | | | |
|---|---|---|---|---|---|---|---|---|---|
| | 1 | 2 | 3 | 4 | 5 | 6 | 7 | 8 | 9 |
| | 9 | 8 | 7 | 6 | 5 | 4 | 3 | 2 | 1 |
| | 1 | 2 | 3 | 4 | 5 | 6 | 7 | 8 | 9 |
| | 9 | 8 | 7 | 6 | 5 | 4 | 3 | 2 | 1 |
| | | | | | | | | | 2 |
| 2 | 2 | 2 | 2 | 2 | 2 | 2 | 2 | 2 | 2 |

# 4

# FASCINATING MATH. OPERATIONS (LOGIC)

**Theorem one —**

**1. Can you prove 2 = 1 (Boolean logic)**

It is a well-known fact that 2 is not equal to 1 but you can prove it as follows:

we know that

$$a^2 - a^2 = (a-a)\ (a+a)$$

$$a\ (a-a) = (a-a)\ (a+a)$$

Dividing both sides by (a–a)

$$a = a+a$$

or $a = 29$

dividing both sides by a

$$1=2$$

or $2=1$

**Theorem two —**

**2. Can you prove 3 = 4**

3 is not equal to 4 but you can surprise your friends by proving it as follows:

We know that $-12 = -12$

This can be written as $9-21=16-28$

$$(3)^2 - 7 \times 3 = 4^2 - 7 \times 4$$

$$(3)^2 - \frac{7}{2} \times 3 \times 2 = (4)^2 - \frac{7}{2} \times 4 \times 2$$

Adding $(\frac{7}{2})^2 =$ on both sides, we have

$$(3)2 - \frac{7}{2} \times 3 \times 2 + (\frac{7}{2})^2 = (4)^2 - \frac{7}{2} \times 4 \times 2 + (\frac{7}{2})^2$$

$$(3-\frac{7}{2})^2 = (4-\frac{7}{2})^2$$

$$\text{or } 3-\frac{7}{2} = 4-\frac{7}{2}$$

or 3

One can rationalise this logic with any of the natural numbers. But in the basic Boolean operation of numbers, an identical element does exist for each operation or each one is distributive over another. To say, for every element in the set, there shall be another which combined with the initial one would yield an identical element of the other operation.

3. **Addition without computation**

Ask someone to write any five-figure number on the black board or paper. Write a five-figure number beneath at random. You could permutate in a manner, that each one added to the digit above will tend to become nine. For example:

His number 45623
Your number 54376

Tell him to put a third five-figure number beneath your number, say 36258. Write a fourth number, 63741, using similar principle yourself. The moment he has finished the fifth number, say 48765 you draw a line underneath and without a moment's hesitation write the correct total.

Amazingly, you could write it from left to right. How will you do it? Subtract 2 from the fifth number and put 2 in front of the remainder. For example, if the fifth number is 48765, the total will be 248763.

4. **Multiplying any number by 9999....**

To multiply a number by 9, 99, 999, 9999 ......... annex as many zero's as there are 9s in the multiplier and from the result subtract the number itself. Thus, $345 \times 99 = 34500 - 345 = 34155$ or $75 \times 9 = 750 - 75 = 675$.

5. **Multiplying any number by 5**

To multiply a number by 5, annex 0 to the number and divide the result by 2.
Thus, $172 \times 5 = 1720 \div 2 = 860$

6. **Multiplying any number by 125**

To multiply a number by 125, annex 000 to the number and divide the result by 8

Thus, $89 \times 125 = 89000 \div 8 = 11125$

7. **Magic multiplication**

Ask someone to write down the numbers 123456789 on a piece of paper. Tell his figures were incompatible, so he will pause over the worst number. Suppose he says 9, then tell him to multiply the number 12345679 by 81.

You will get 999999999.

Suppose his number is 4 then you would multiply the line by 36 and your answer would be 444444444. Whatever number your pal puts in, you should multiply it by 9 and you shall find all those articulated by him. This is the rationalised approach for such a random selection.

8. **Multiplying any number by 11**

Do you need any number be multiplied by 11, start from right to left and write the first digit per se. Add the first and second digits and write the unit placed number (if the sum is greater than 9, then carry over the remainder to the sum of the second and third digits). Write the unit placed integer of the sums and continue so. Finally write the last digit as it is. For example $345 \times 11$, first write 5, now add 5 and 4 and write 9, then add 4 and 3 and write 7. Finally write 3. So $345 \times 11 = 3795$. Similarly, $8165 \times 11 = 89815$ and $397842 \times 11 = 4376262$.

9. **Squaring any integer of nine**

If you want the square of a number which contains only 9s, write from right to left first 1, then zeros, one less than the number of 9s, then write 8. Finally write as many 9s as the zeros e.g. $999^2 = 998001$. Similarly, $999999^2 = 999998000001$.

10. **Squaring any integer ending in 5**

If you want to square a number whose last digit is 5 e.g. 35, 45, etc., begin from right to left — first write $5^2$ i.e. 25, then multiply the other digit of number to its next consecutive number (i.e. number+1) e.g. $35^2 = 35 \times 35 = 1225$. Initially, we have written 25, then 12 which is $3 \times 4$ or $3 \times (3+1)$. Similarly $85^2 = 85 \times 85 = 7225$ and $45^2 = 45 \times 45 = 2025$.

11. **Adding consecutive numbers**

If you want to add consecutive integers, add the first and the last integers. Then multiply the result by half the number of integers: e.g. $14+15+16+17+18+19=(14+19) \times \frac{6}{2} = 99$.

As the above series has six numbers. Similarly $40+41+42+43+$ $\ldots\ldots\ldots+49=(40+49) \times \frac{10}{2} = 89 \times 5 = 445$.

12. **Sum of consecutive odd numbers**

From 1 to 100 there are 50 odd numbers i.e. $1+3+5+7 \ldots\ldots +99 = 50 \times 50 = 2500$

Also $1+3+5+7+9+11+13=49$. They are in an arithmetic series of odd numbers, with a constant 2.

13. **Sum of consecutive even numbers**

We could multiply the numbers in the group by one more than their number, e.g. the sum of all even numbers from 1 to 100, i.e. $2+4+8 \ldots\ldots\ldots +100 = 50 \times 51 = 2550$.

Also $2+4+6+8+10+12+14+16=8 \times 9=72$. They are in an arithmetic progression of even numbers with a constant 2.

14. **Multiplying any number by 25**

To multiply a number by 25, we move the decimal point of the given number two places to the right and divide by four, e.g.,: $53219 \times 25 = \frac{5321900}{4} = 1330475$.

Rationalising we find if x is multiplied by 25 and divided by 4 we will get $\frac{25}{4} \times = 6.25 \times$ or $\frac{1}{8}$ x.

15. **Squaring any two-digit number, e.g., 86:**

*Step 1* — Square the unit digit of the number $(6)=36$, we write 6 on the extreme right and carry over 3.

*Step 2* — Double the product of individual numerals and add the earlier carry-over i.e. $2 \times 8 \times 6 = 96 + 3 = 99$. Write 9 to the left of unit digit 6. That is 96 and carry over 9.

*Step 3* — Square tens digit and add carry-over and place the sum to the extreme left of the number written earlier

i.e. $(8)^2 = 64 + 9 = 73$ therefore

$(86)^2 = 7396$

**Example — Square 53**

(1) $3 \times 3 = 9$ 9

(2) $2 \times 5 \times 3 = 30$ 09
carry over 3

(3) $5 \times 5 = 25 + 3 = 28$

Therefor square of 53 = 2809

Rationalising, we fill find, if xy is the two digit number, $y^2$ (in the first stage), product inversion as 0.2yx (in the second stage) or o.yz and $x^2= x^2+y=[x^2+y+0y.x]$ (in the third stage), remain the squared product.

## 16. Tests of divisibility

A given number is divisible by

2 — If the unit digit of the number is even or zero

3 — If the sum of the digits of the given number is dividable by 3

4 — If the number formed by the last two digits is evenly divisible by 4

5 — If the last digit is either 0 or 5

6 — If the sum of the digits of any even number is divisible by 3

8 — If the number formed by the last three digits of the given number is divisible by 8

9 — When the sum of the digits of the given number is equal to 9 or multiple of 9.

11 — Add alternate digits to form two sums. If the difference between the two sums is equal to zero or a multiple of eleven, the given number is evenly divisible by 11.
854037624
Sum of alternative digits starting from unit digits = 4 + 6 + 3 + 4 + 8 = 25
Sum of alternate digits starting from ten = 2 + 7 + 0 + 5 = 14
The difference of two sums = 25 - 14 = 11 which is a multiple of 11. Hence the number is divisible by 11.

## 17. The cube roots

Generally students face difficulty in finding the cube roots of the numbers. The cube roots are found either by logarithmic method or simple factorisation.

One can find the cube roots of any number upto six digits. Only numbers which are perfect cube can be found out. The cube roots of numbers of six digits will be in 2 digits i.e. in unit digit and the ten's digit places. The following tables are used to find the cube roots. For unit digit see table A.

**Table A** (For unit digit)

Unit place in the given number 0, 1, 2, 3,4, 5, 6, 7, 8, 9

Unit digit in the answer 0,1, 8, 7, 4, 5, 6, 3, 2, 9

Check the last three digits leaving second and third digits of the given number with Table B

**Table B** (For ten's digit)

Number with last three digits 1, 8, 27, 64, 125, 216, 343, 512, 729

Ten's digit place in the answer 1, 2, 3, 4, 5, 6, 7, 8, 9

Some of the examples are:

**a) Cube root of $\overline{512}$**

The unit digit in this number is 2. Looking at table A, it is evident that unit place is 8.

In the given number three is no number in 4th, 5th, 6th place, hence the answer is only in the digit. So the cube root of 512 is 8.

**b) Find the cube root of $\overline{1728}$**

The unit digit of the number is 8, so by looking in Table A we get the unit place digit as 2. As the number 1 is in the 4th place and no number is there in 5th and 6th places so from Table B.

The 10th place is 1.

Hence the cube root of 1728 is 12.

**c) Find the cube root of $\overline{32}\,\overline{768}$**

As the unit digit of number is 8, hence from table A the unit digit is 2. 3 and 2 are on 5th and 4th place, respectively so from Table B the tenth place digit corresponding to 32 is 3. Hence the cube root of the given number is 32.

**d) Find the cube root of $\overline{753}\,\overline{571}$**

For unit digit 1 the number corresponding to unit place is 1 (from Table A). The tenth place digit corresponding to 753 is 9 from Table B. Hence the cube root of 753571 is 91.

By using this method frequently you will get familiar with this method and will prove to be an important tool in your life.

## 18. Repeated numbers

There are many occasions when repeated numbers behave in a similar fashion, while they are in arithmetic progression. The approach hardly applies to the numbers in harmonic series.

**Example 1** When number 111 is multiplied by 12, 23, 34, etc.

i.e. in increasing order of 11 than the answer differs from previous one only by 1221. This you can see as follows:

$111 \times 12 = 1332$
$111 \times 23 = 2553$ — Difference = 1221

$111 \times 34 = 3774$
$111 \times 45 = 4995$ — Difference = 1221

**Example 2** If you take squares of 11, 111, 1111, etc. then do you know that the answer contains odd digits, and also the square always starts with one and gradually increases to a maximum at the middle and decreases in the same way it has increased. This is shown below:

$11 \times 11 = 121$

$111 \times 111 = 12321$

$1111 \times 1111 = 1234321$

**Example 3** If we multiply 11 by repeated digits of 2 such as 2, 22, 222, 2222, etc. then the beauty of the answer lies in the fact that the first and last digits are 2 and all other digits are 4. This is shown below:

$11 \times 2 = 22$

$11 \times 22 = 242$

$11 \times 222 = 2442$

$11 \times 2222 = 24442$

## 19. A useful way of multiplication

You can multiply two numbers of two digits each when the ten's places are the same and the sum of unit digits is ten. You have to do the following.

Multiply the ten's place digit by the number one bigger and multiply the two unit placed digits together. Write the two answers in apposition reading from left to right. You will find the result of your multiplication.

**Example 1** $73 \times 77$
$7 \times 8 = 56$
$3 \times 7 = 21$
Answer is 5621

**Example 2** $92 \times 98$
$9 \times 10 = 90$
$2 \times 8 = 16$
Answer is 9016.

**Example 3** $88 \times 82$
$8 \times 9 = 72$
$8 \times 2 = 16$
Answer is 7216.

■■

# 5

# FUN WITH FIGURES

**Fun 1 — How many rectangles are there in the Figure 5.1?**

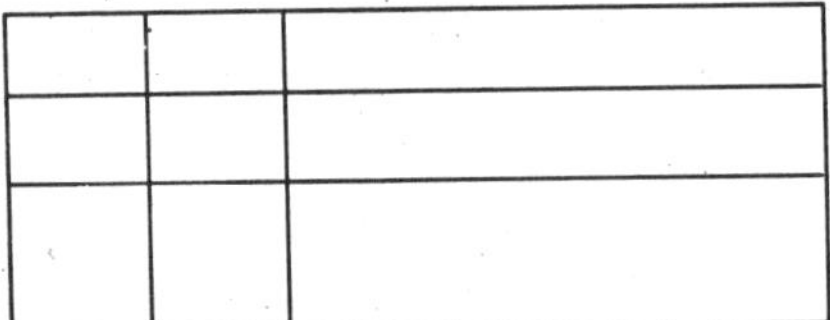

*Fig. 5.1*

**Fun 2 — How many triangles are there in Figure 5.2?**

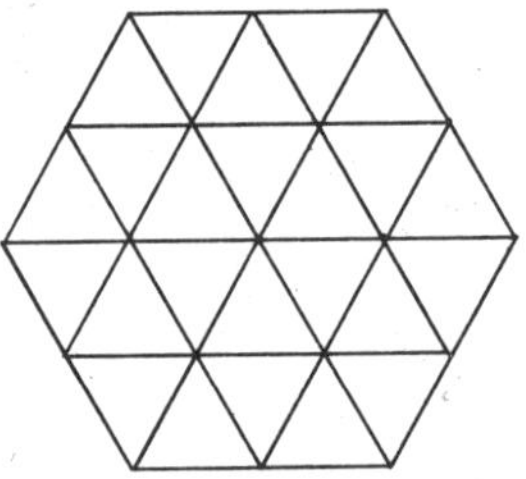

*Fig. 5.2*

**Fun 3 — How many triangles are there in Figure 5.3?**

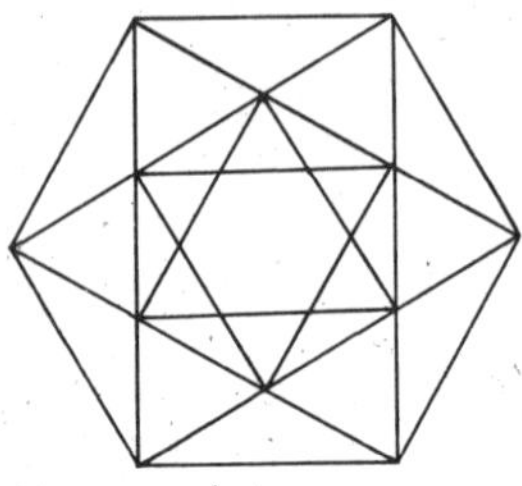

*Fig. 5.3*

**Fun 4 — How many regular hexagons Figure 5.4 contains?**

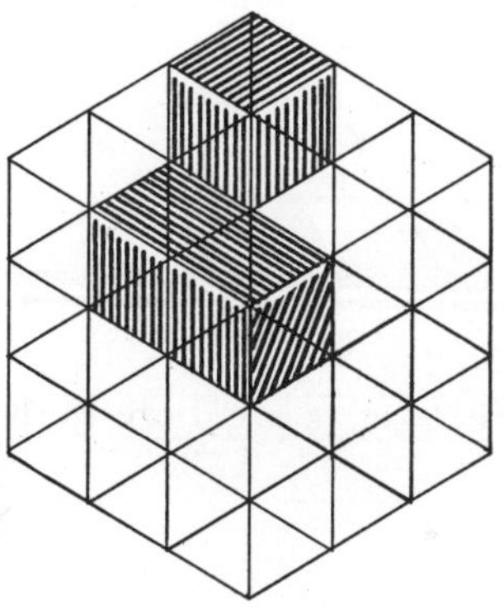

*Fig. 5.4*

**Fun 5 — How many squares are there in Figure 5.5?**

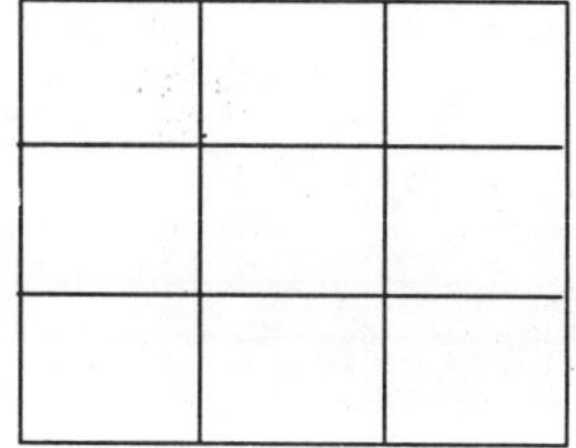

*Fig. 5.5*

**Fun 6 — How many squares are there in Figure 5.6?**

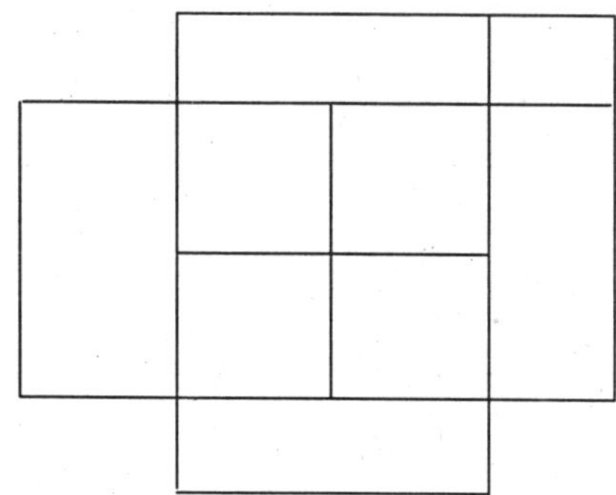

*Fig. 5.6*

**Fun 7 — How many triangles are there in Figure 5.7?**

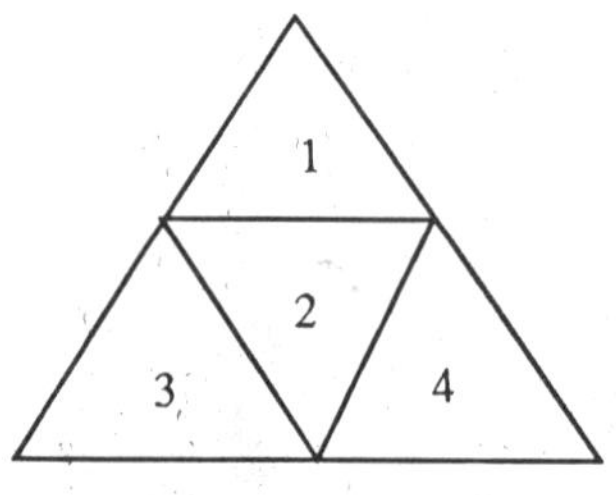


*Fig. 5.7*

**Fun 8 — How many triangles Figure 5.8 contains?**

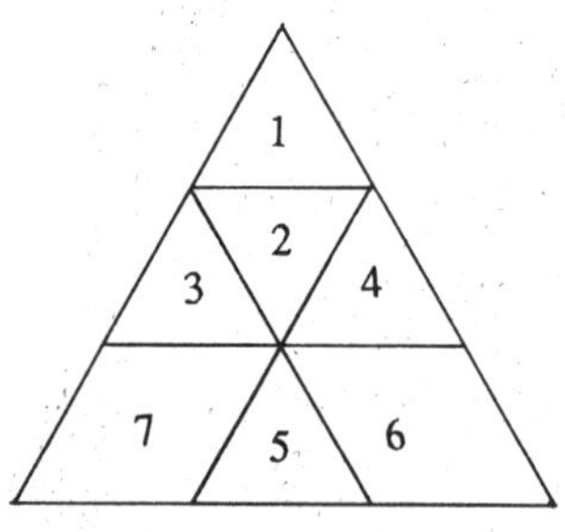


*Fig. 5.8*

**Fun 9 — Can you tell how many triangles are there in Figure 5.9?**

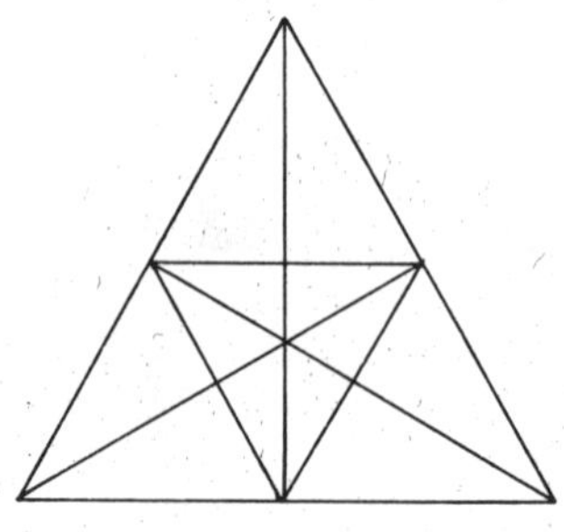

*Fig. 5.9*

**Fun 10**—**How many triangles are there in the star represented by Figure 5.10?**

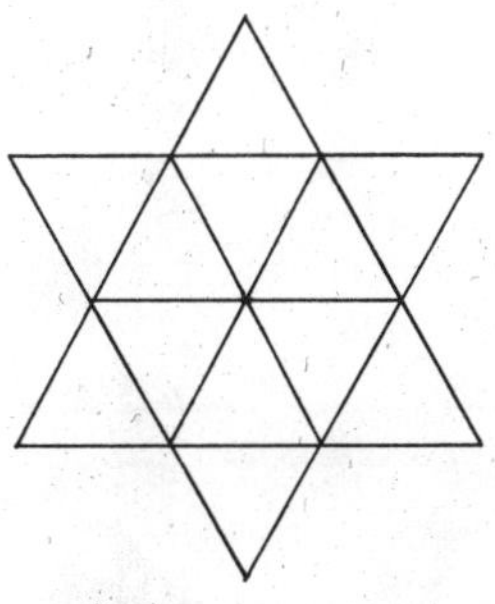

*Fig. 5.10*

**Fun 11**—**How many triangles are there in Figure 5.11?**

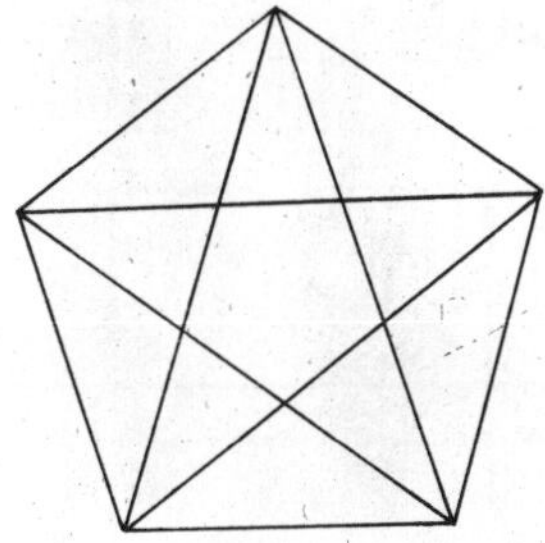

*Fig. 5.11*

**Fun 12**—**How many triangles are there in the Figure 5.12?**

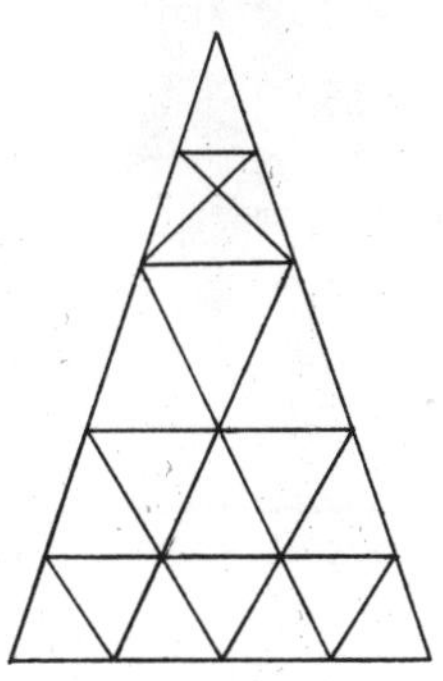

*Fig. 5.12*

**Fun 13—How many cubes are there in Figure 5.13?**

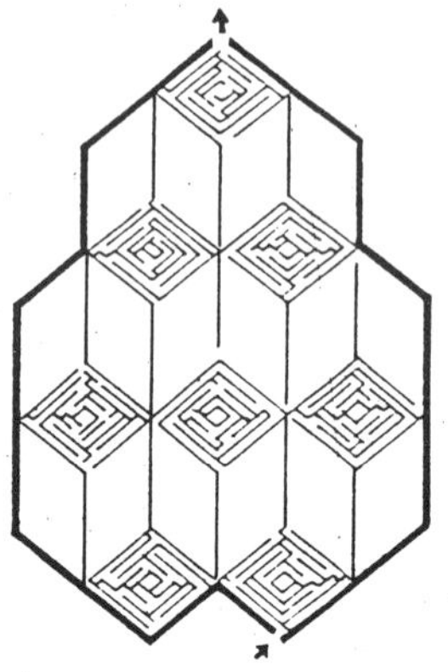

*Fig. 5.13*

**Fun 14—Arrange 25 mangoes in 12 rows so that each row contains 4 mangoes?**

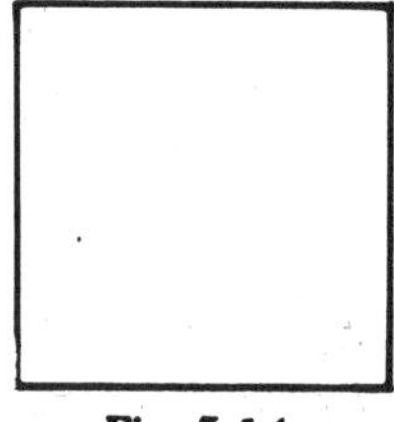

*Fig. 5.14*

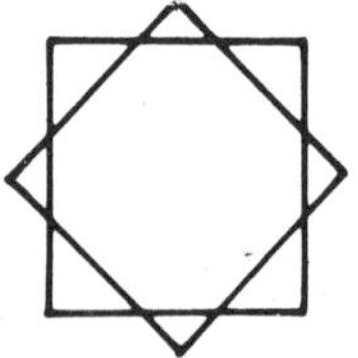

*Fig. 5.15*

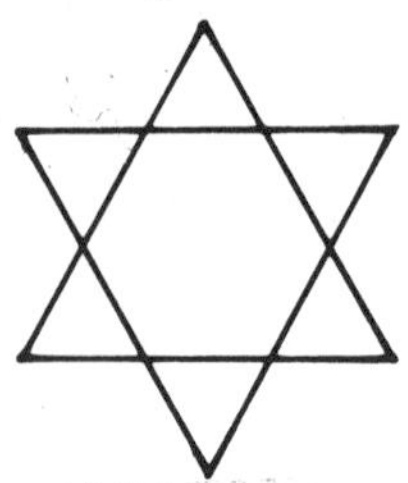

*Fig. 5.16*

# 6

# FUN WITH DOTS

**Fun 1 — Given below are 9 dots. Join these by straight lines to get the maximum number of squares. How many squares can you make?**

**Fun 2 — Given below are 16 dots. Join them in any form to get maximum number of squares.**

**Fun 3 — Given below are 8 dots. How many rectangles can you make?**

Fun 4 — Given below are 13 dots. Join them so as to get maximum number or rectangles.

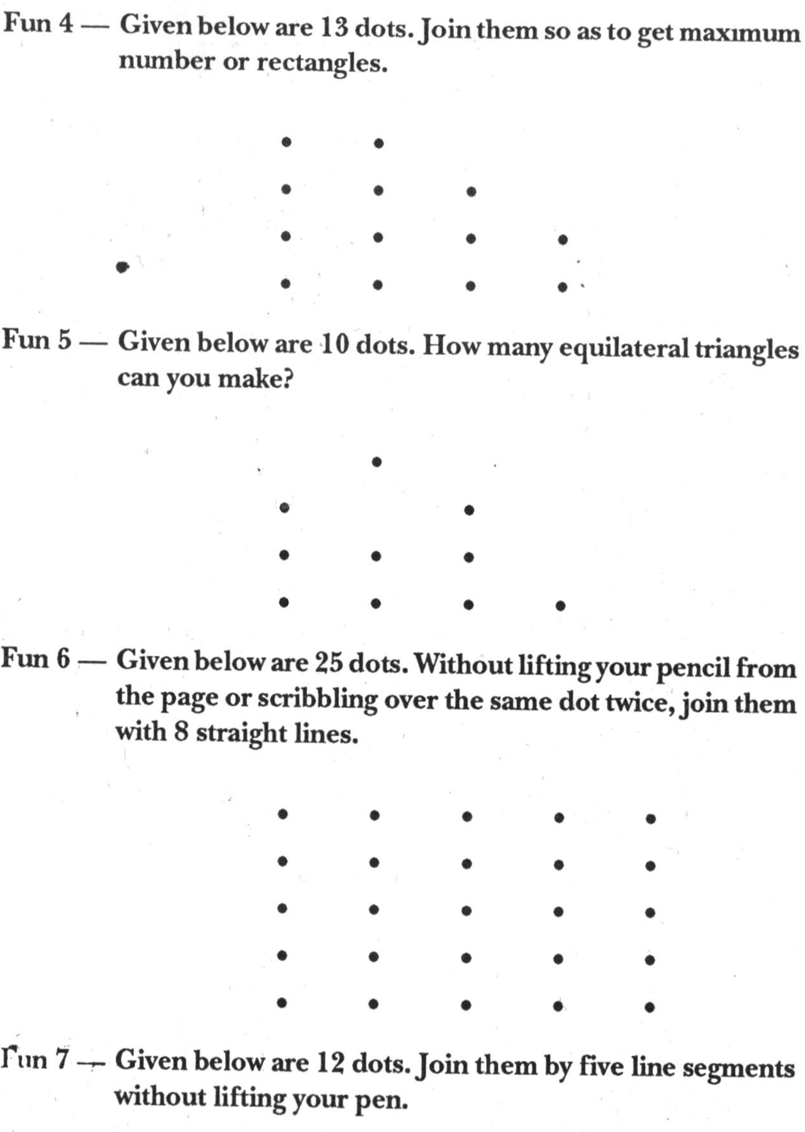

Fun 5 — Given below are 10 dots. How many equilateral triangles can you make?

Fun 6 — Given below are 25 dots. Without lifting your pencil from the page or scribbling over the same dot twice, join them with 8 straight lines.

Fun 7 — Given below are 12 dots. Join them by five line segments without lifting your pen.

■■

# 7

# MATCHSTICK NUMBER FUN

**Fun 1 — Put six matches as shown in Figure below. Now add five more matches to make nine.**

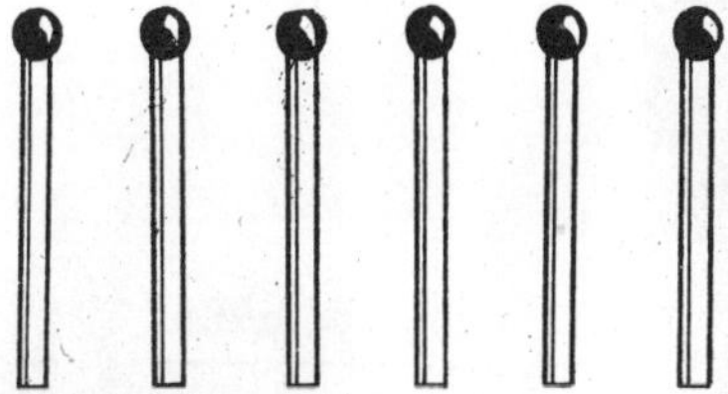

**Fun 2 — Given below is a sum in Roman alphabets. Move one match so that the sum works out.**

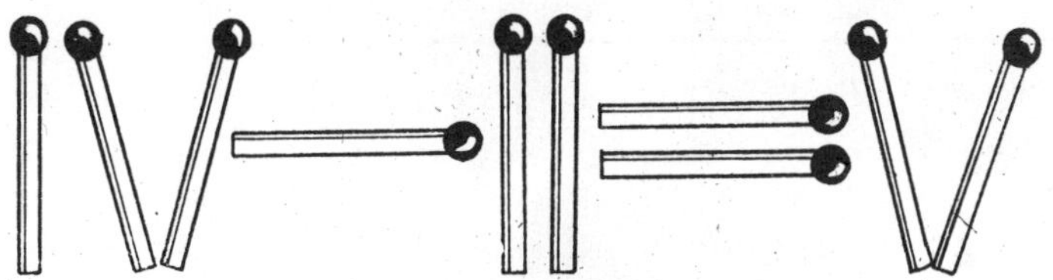

**Fun 3 — Given below is a sum in Roman alphabets. Add one match to make the sum correct, if you concentrate upon.**

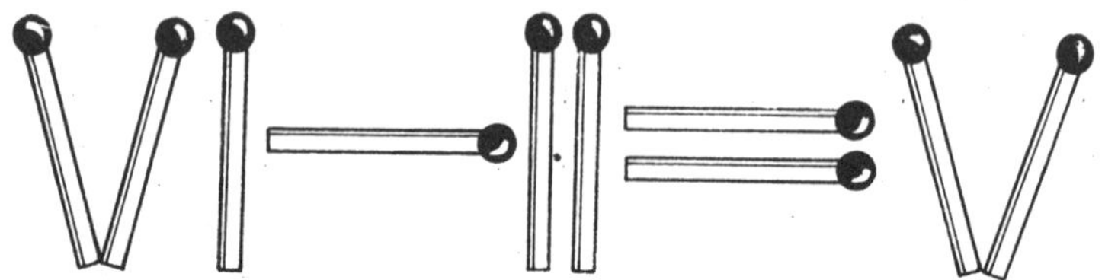

**Fun 4 —** **In the figure below, move two matches to make the sum compatible.**

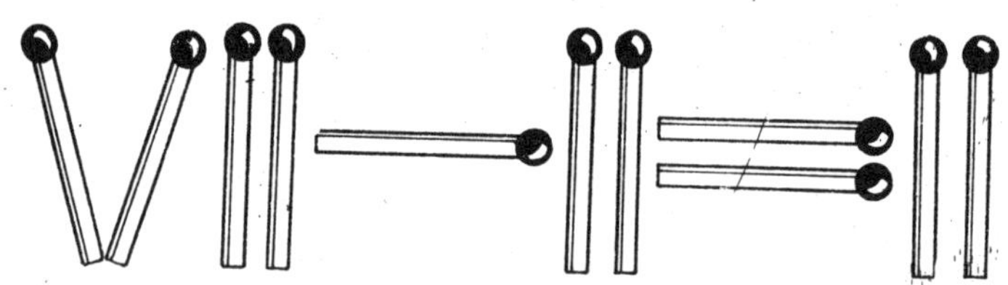

**Fun 5 —** **In the figure below, remove 3 matches to make the sum additive.**

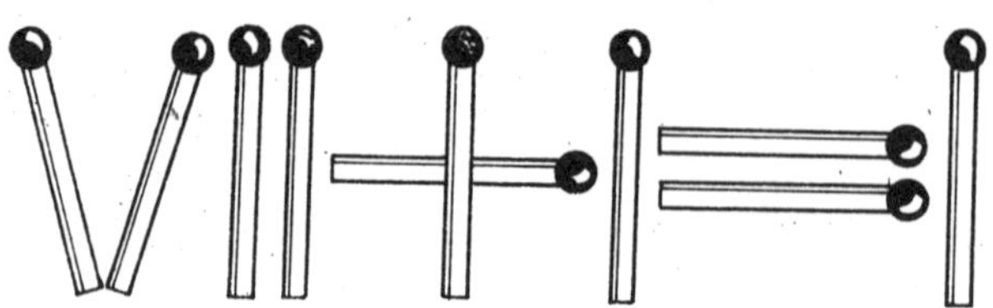

Fun 6 — Given below is a sum in Roman alphabets. Make the sum correct without adding or removing any matches.

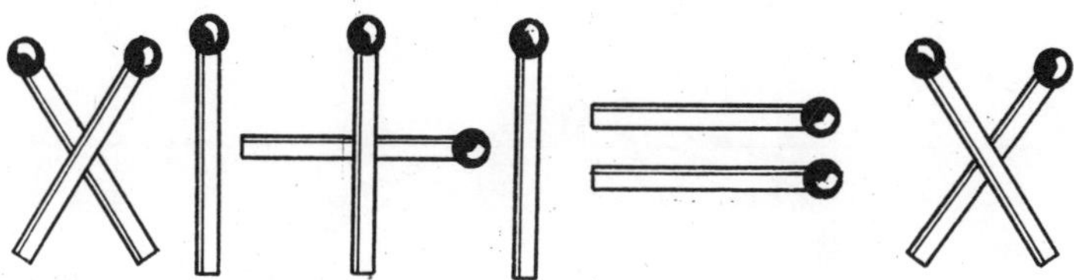

Fun 7 — In the sum given below, move one match to make it correct.

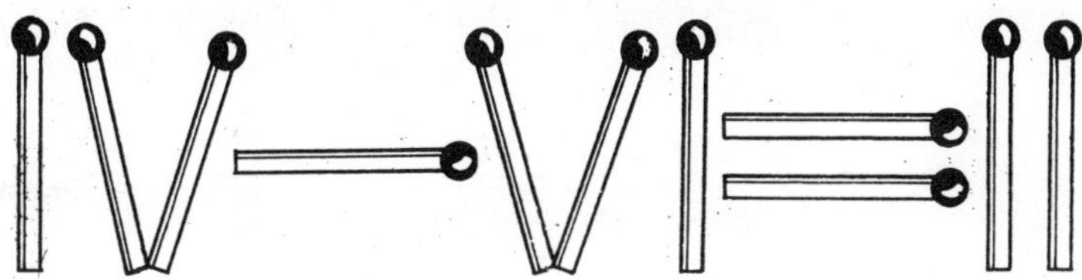

Fun 8 — In the figure below, move just one stick to make the sum at par.

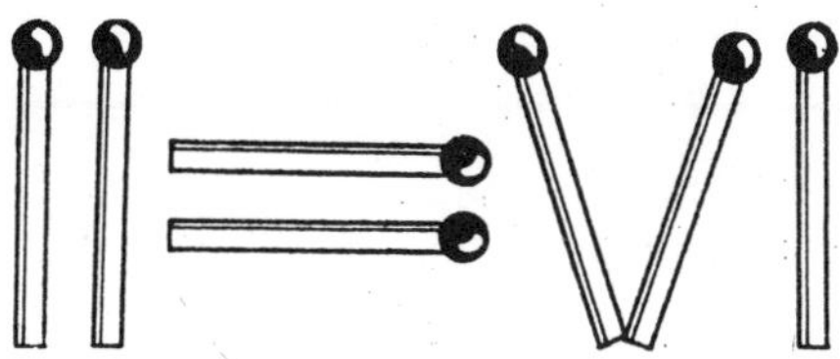

**Fun 9** — Lattice six matches as shown below. Move 3 of them to make 6 equilateral triangles in a distributive phase.

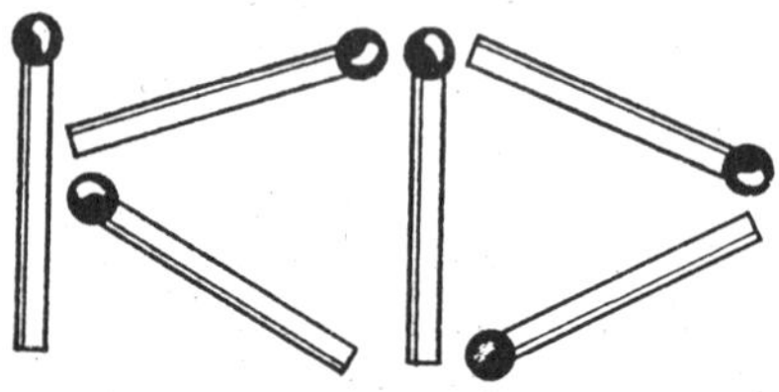

**Fun 10**—In the figure below, 12 matches make six equilateral triangles. Move four of them to make three equilateral triangles.

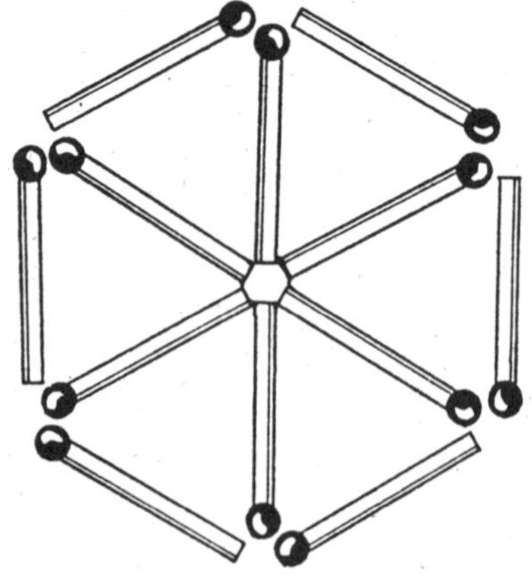

**Fun 11**—How many triangles can you scoop in with 17 matches. Let each side of the triangle carry whole matches.

**Fun 12**—Using 12 match sticks, can you construct a plane figure containing an area of exactly three square match stick lengths? ■■

# 8

# MAGIC SQUARES

The fun of the magic squares is that whichever way you add up the numbers in the square — horizontally or vertically or diagonally — the sum always remains the same. The squares may contain nine, sixteen, twenty-five, thirty-six or forty-nine boxes. Here are two examples:

| | | |
|---|---|---|
| 10 | 3 | 8 |
| 5 | 7 | 9 |
| 6 | 11 | 4 |

(a)

| | | | |
|---|---|---|---|
| 23 | 10 | 9 | 20 |
| 12 | 17 | 18 | 15 |
| 16 | 13 | 14 | 19 |
| 11 | 22 | 21 | 8 |

(b)

The answer of the magic square (a) is 21 and that of (b) is 62, i.e. sum of each row, and column and diagonal is same. Two standard squares of 9 and 16 boxes are given below :

| | | |
|---|---|---|
| 8 | 1 | 6 |
| 3 | 5 | 7 |
| 4 | 9 | 2 |

| | | | |
|---|---|---|---|
| 16 | 2 | 3 | 13 |
| 5 | 11 | 10 | 8 |
| 9 | 7 | 6 | 12 |
| 4 | 14 | 15 | 1 |

These two are used for making other magic squares. Magic squares are of two types:

(i) Magic squares containing odd number of boxes such as 9, 25, 49, etc.

(ii) Magic squares containing even number of boxes such as 16, 36, 64, etc.

**Legend of magic squares**

An old Chinese legend sayeth, many centuries ago, a turtle crawled out of Yellow River. On its back were strange designs formed by dots. People formed a magic square (Fig. 8.1) by aping down the number of dots. Its sum perhaps remained same irrespective of addition in horizontal, vertical or diagonal modes.

People thereupon devised magic squares at variance, following the curiosity behind.

**(i) Magic squares with odd number of boxes**

We wish to construct a magic square of 9 boxes in which the sum of each row, column and diagonal shall be 36. Diagonally the sum remained at 36. It contained three rows and three columns so the number in the central box would be 36/3=12.

| | | |
|---|---|---|
| | | |
| | 12 | |
| | | |

As this magic square has nine boxes, we have to consider nine numbers, including 12. These numbers are 8, 9, 10, 11, 12, 13,

14, 15, 16 in A.P. with a constant **one.** We make sets using these numbers in a way as to arrive upon a sum of 36. Different sets of permutations are given below :

8+11+16 = 36 (i)
9+12+15 = 36 (ii)
10+12+14 = 36 (iii)
11+12+13 = 36 (iv)

12 is the common factor from (ii) to (iv). The arrangement of numbers diagonally leads to —

| | | |
|---|---|---|
| 9 | | 11 |
| | 12 | |
| 13 | | 15 |

(A) Diagonally

For first row, two numbers are 9 and 11, which add up to give 20. The third number is chosen as to give a total of 36 i.e. 9+11+16=36. Therefore the number is 16.

For the third row, two numbers 13 and 15 make 28, so the third number has to be 8 in order to give a total of 36. The sum of numbers of first and third columns are 22 and 26 respectively hence the other numbers are 14 and 10.

So our magic square is:

| | | |
|---|---|---|
| 9 | 16 | 11 |
| 14 | 12 | 10 |
| 13 | 8 | 15 |

This is the magic square which gives a total of 36 when added in any mode, viz. — vertically, horizontally or diägonally.

**(ii) Magic squares with even number of boxes**

The method used here is little difficult and differs from that applied for odd number of boxes.

If we want to construct a magic square containing 16 boxes with the help of numbers from 1 to 16, then first write all the numbers in the blank squares as given below:

| 1 | 2 | 3 | 4 |
|---|---|---|---|
| 5 | 6 | 7 | 8 |
| 9 | 10 | 11 | 12 |
| 13 | 14 | 15 | 16 |

If you change the order of numbers in the diagonals i.e. change 1, 6, 11, 16 into 16, 11, 6, and 4, 7, 10, 13 into 13, 10, 7, 4 you will end up with a magic square in which sum of each row, each column and diagonals is 34.

| 16 | 2 | 3 | 13 |
|---|---|---|---|
| 5 | 11 | 10 | 8 |
| 9 | 7 | 6 | 12 |
| 4 | 14 | 15 | 1 |

*P.S*: If any constant number be multiplied, divided, added or subtracted to each number of a given magic square it will produce another magic square.

**Construction upon set norms:**

One can construct a magic square when any specific condition is given. For example, if we want to construct a magic square of nine boxes with its horizontal rows, vertical columns and diagonals giving 42 on addition and the number 15 on any of the boxes as shown in next page:

| | | |
|---|---|---|
| | | 15 |
| | | |
| | | |

The standard magic square with 9 boxes is:

| | | |
|---|---|---|
| 8 | 1 | 6 |
| 3 | 5 | 7 |
| 4 | 9 | 2 |

As 15 is given in that box which contains 6 in the standard magic square.

In terms of 6, one can write 15 as:

(i) $15 = 6 + 9$

(ii) $15 = 6 \times 3 - 3$

We can have the following magic square:

(i)

| | | |
|---|---|---|
| 8 + 9 | 1 + 9 | 6 + 9 |
| 3 + 9 | 5 + 9 | 7 + 9 |
| 4 + 9 | 9 + 9 | 2 + 9 |

=

| | | |
|---|---|---|
| 17 | 10 | 15 |
| 12 | 14 | 16 |
| 13 | 18 | 11 |

That is 9 is added in each box. You can infer that its magic constant is 42.

(ii) If we multiply each box by 3 and subtract 3, the resultant magic square shall have a magic constant of 36.

| 8 × 3–3 | 1 × 3–3 | 6 × 3–3 |
|---|---|---|
| 3 × 3–3 | 5 × 3–3 | 7 × 3–3 |
| 4 × 3–3 | 9 × 3-3 | 2 × 3–3 |

=

| 21 | 0 | 15 |
|---|---|---|
| 6 | 12 | 18 |
| 9 | 24 | 3 |

Similarly, those with 16 or more boxes can be constructed if some condition is given just by comparing with the standard magic square.

In this example, a magic square with 16 boxes is drawn according to set initial conditions as shown in figure below.

| | | | |
|---|---|---|---|
| 25 | | | |
| | | | 32 |
| | | | |

Standard square is :

| 16 | 2 | 3 | 13 |
|---|---|---|---|
| 5 | 11 | 10 | 8 |
| 9 | 7 | 6 | 12 |
| 4 | 14 | 15 | 1 |

Adding 20 to each box, we get the modified square which satisfies these conditions as :

| 36 | 22 | 23 | 33 |
|---|---|---|---|
| 25 | 31 | 30 | 28 |
| 29 | 27 | 26 | 32 |
| 24 | 34 | 35 | 21 |

Its magic constant is 114.

**Magic square 1**

Make a magic square of magic constant 105 with 25 boxes. Some numbers are given below:

| | | | | |
|---|---|---|---|---|
| 25 | | | | 23 |
| | | | | |
| | | 21 | | |
| | | | | |
| 19 | | | | 17 |

**Magic square 2**

Carve out a magic square of 16 boxes with magic constant of 150. Some box numbers are given below:

| | | | |
|---|---|---|---|
| | | 31 | |
| 34 | | | |
| | 35 | | |
| | | | 30 |

**Magic square 3**

Draw a magic square of 16 boxes with magic constant 50. Digits are to be used from 5 to 20.

**Magic square 4**

Outline a magic square of 16 boxes with magic constant 34 and having some fixed numbers as shown below:

| | | | |
|---|---|---|---|
| 1 | | | 4 |
| | | | |
| | | 6 | |
| 13 | | | |

**Magic square 5**

Draw a magic square of 5 × 5 with magic constant of 165 using the numbers from 21 to 45.

**Nasik magic square**

Given below is a 5 × 5 magic square. Fill it with numbers 1 through 25 in a way that the sum of the numbers in each row, each column and each of the two diagonals stretch on to 65. The number 1 should be at the Central Square.

# 9

# NUMBERING OF FIGURES

1. **Magic star**

   In Figure 9.1, fill the circles with numbers from 1 to 12 in a way that the sum of each row is 26. No number should be written twice:

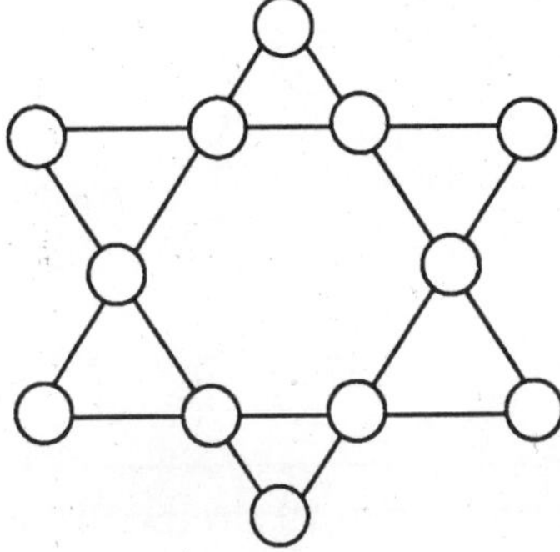

*Fig. 9.1*

2. **Honey comb**

   Below you will see a honey comb (Fig. 9.2). Place the numbers from 1 to 19 in each cell in a way that there is a difference of at least 4 between any one cell and its neighbouring oncs.

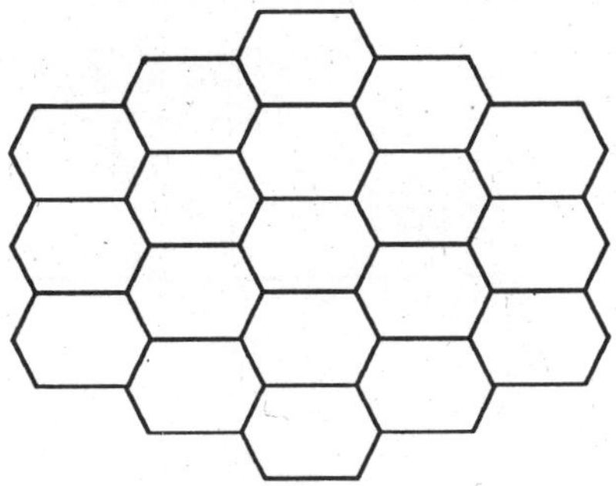

*Fig. 9.2*

3. **Magic figure**

Fill the circles in Figure 9.3 with integers from 1 to 11 in a way that accounting any three directly connected circles and adding up their numbers, all numbers between 6 and 30 can be formed.

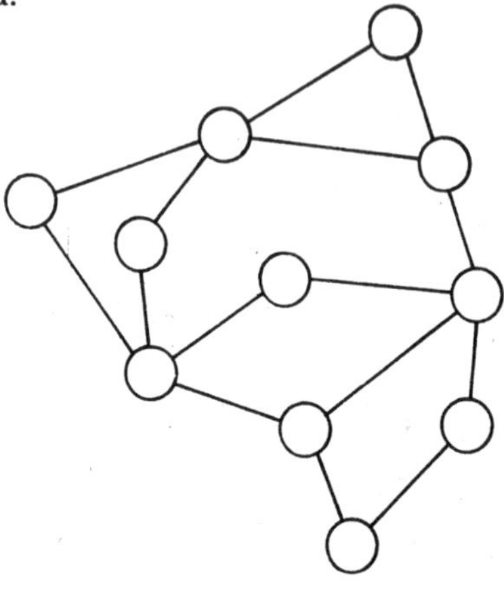

*Fig. 9.3*

4. **Adding wheel**

Here is a small wheel with 19 circles. Place the numbers from 1 to 19 in the circles so that every straight line of 3 circles add up to 30 (Fig. 9.4).

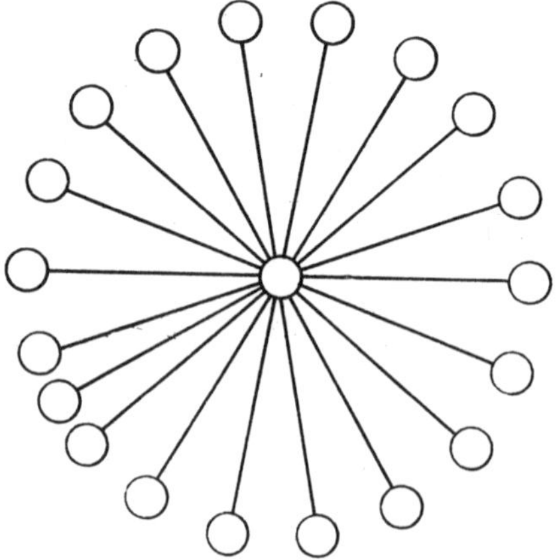

*Fig. 9.4*

5. **Magic circles**

Arrange numbers from 1 to 11 in the circles of the given figure 9.5 so that the sum of the numbers along any straight line shall remain same.

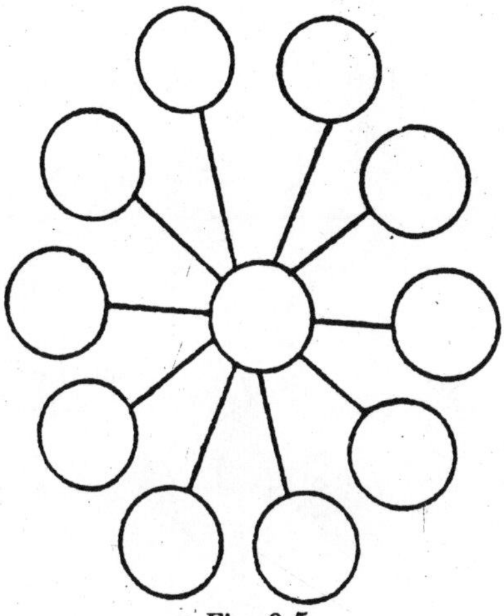

*Fig. 9.5*

6. **Number wheel**

Here is a number wheel or cog wheel (Fig. 9.6). Insert numbers from 1 to 9 in the circles, so that the sum of the three numbers on each line is 15.

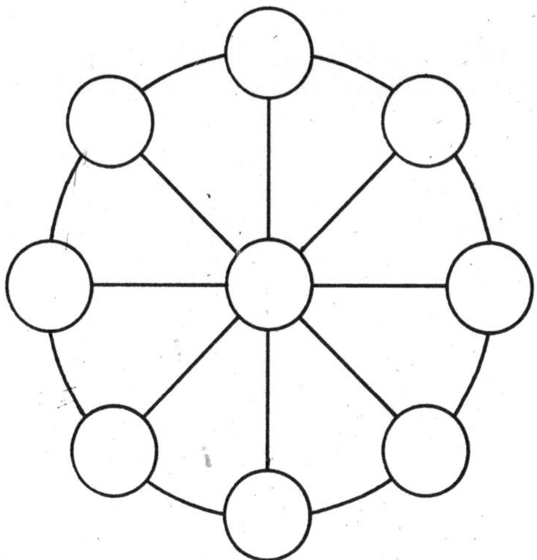

*Fig. 9.6*

7. **Eight-point star**
In an eight-point star (Fig. 9.7), fill the circles with numbers from 1 to 16 so that each line would stretch over to 34.

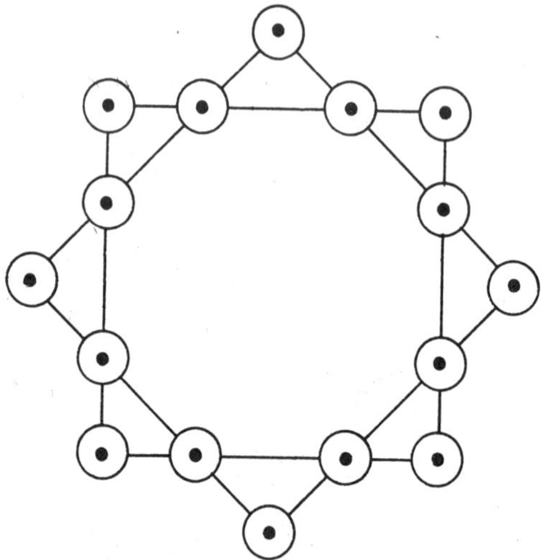

*Fig. 9.7*

8. **The magic hexagon**
Given below (Fig. 9.8) is a hexagon with 19 circles. Fill each circle with first 19 natural numbers in a way that triplet of them in each side or in each of the six radial lines would bring in 23. It can be tried with two permutations.

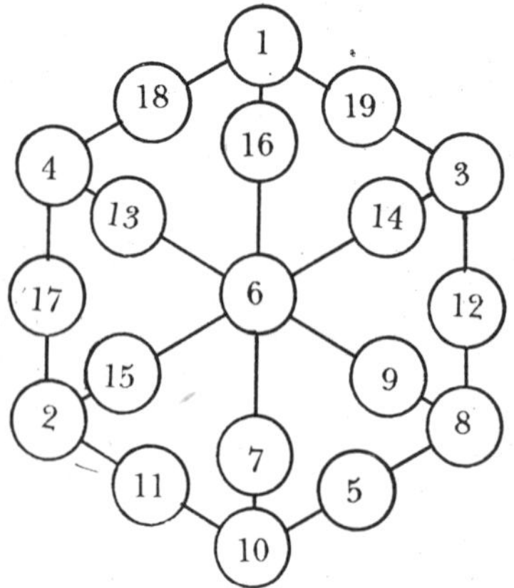


*Fig. 9.8*

## 9. Missing number

Given below is a 3 × 3 magic square in which one number is missing. Can you tell the missing number (Fig. 9.9).

| 8 | 15 | 10 |
|---|---|---|
| 13 | 11 | 9 |
| 12 | 7 | ? |

*Fig. 9.9*

## 10. Penta star

In the given star (Fig. 9.10), five circles have been assigned some numbers but other five circles are empty. Put five different numbers in these circles so that the sum of the four numbers in each line of the figure shall remain the same.

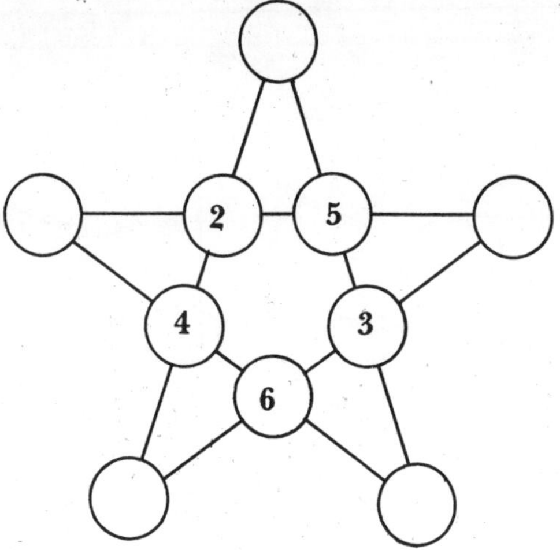


*Fig. 9.10*

## 11. Magic circle

In the circular figure 9.11 below, two digits are missing. They follow a definite sequence. Insert the missing ones.

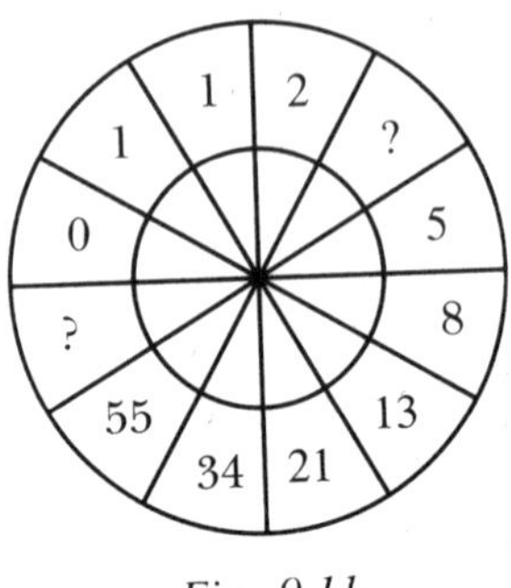


*Fig. 9.11*

12. **Magic movement**

Here are two blocks A and B (Fig. 9.12) each containing four digits. Move only one block so that the sums of numbers in each shall remain the same.

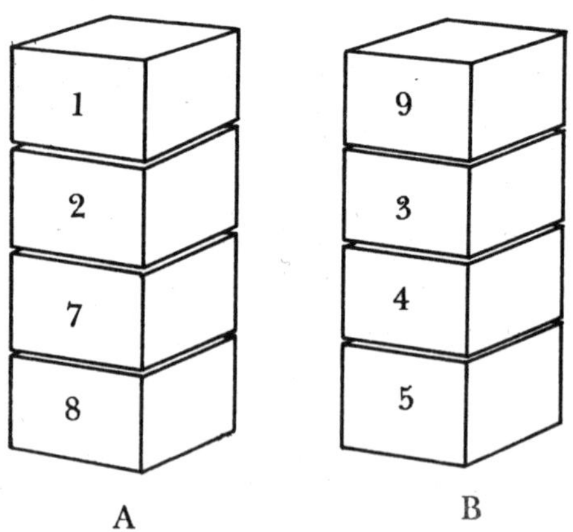


*Fig. 9.12*

13. **Strange star**
Fill the magic star given below with numbers so that each of the 5 lines add up to 24. No number should be repeated.

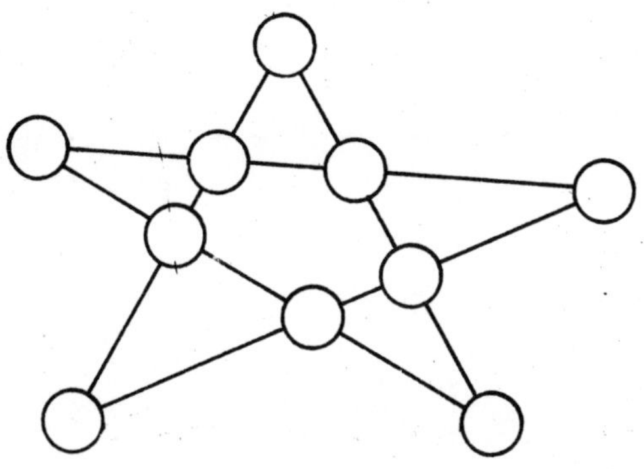

*Fig. 9.13*

14. **Lollipop game**
Six lollipops are arranged as in Figure 9.14. How would you write digits from 1 to 7 in a way so that each row with three digits adds up to 12.

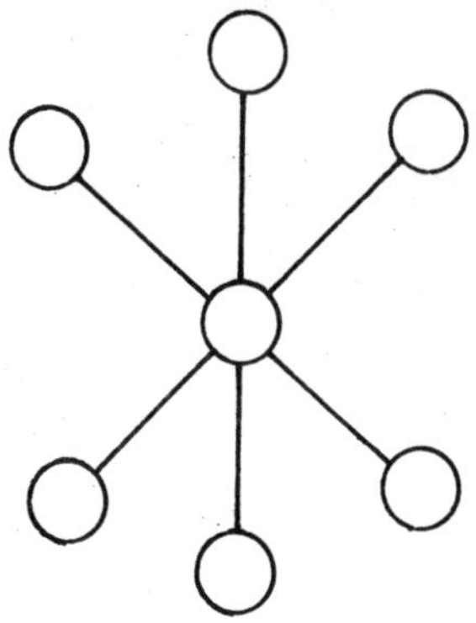

*Fig. 9.14*

15. **Seven point star**

Fill the digits 1 to 14 in the seven-point star in such a way that digits in the four circles of each of the eight rows add up to 30.

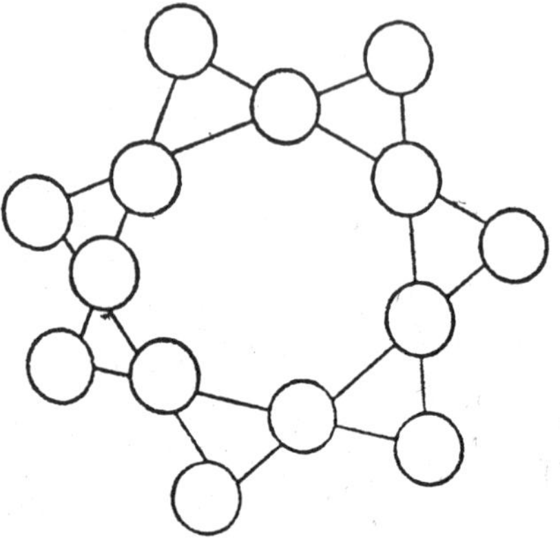

*Fig. 9.15*

16. **Eight point star**

How would you put the digits from 1 to 16 in the circles of ar eight point star so that each line of four circles adds up to 34'

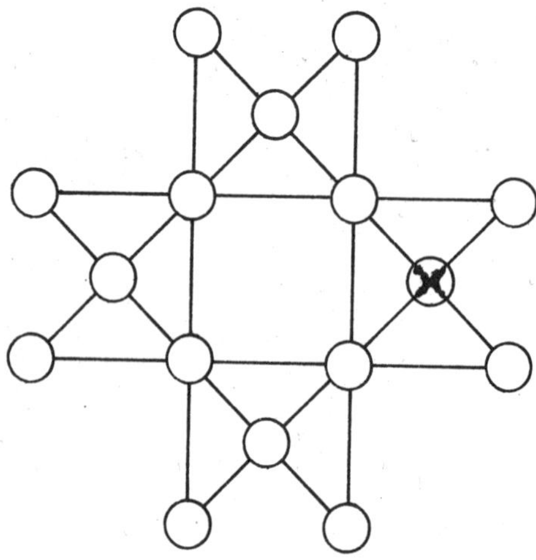

*Fig. 9.16*

# 10 TEST YOUR JUDGEMENT

**Longer line**

How does the length of straight lines in A and B be compared?

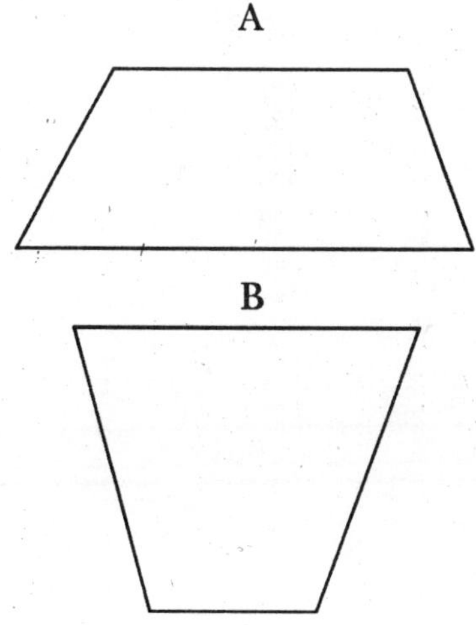


**Dark and bright**

Which circle is larger in magnitude — white or black?

**Two perpendicular lines**

Which line is longer, vertical or horizontal?

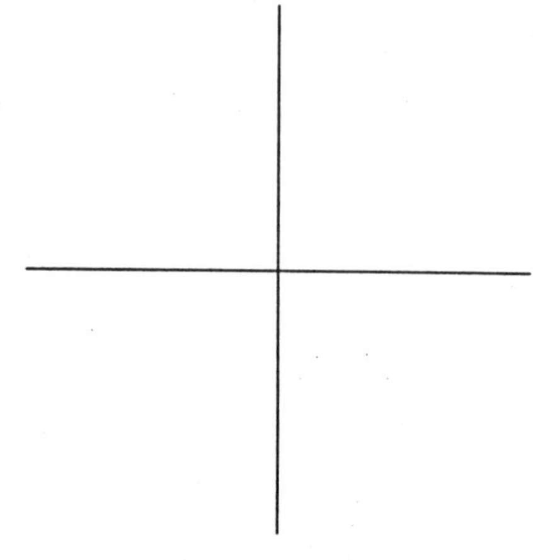

**Longest line**

Which of the three straight lines a, b and c is the longest?

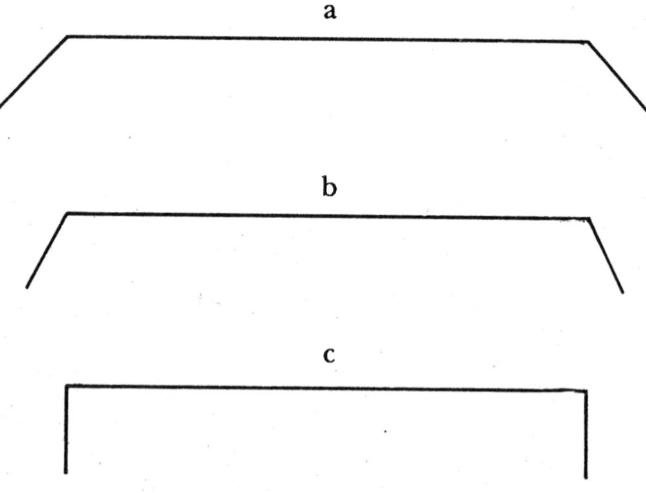

**Intersecting lines**

Which of the lines b or c is aligned with oblique line a?

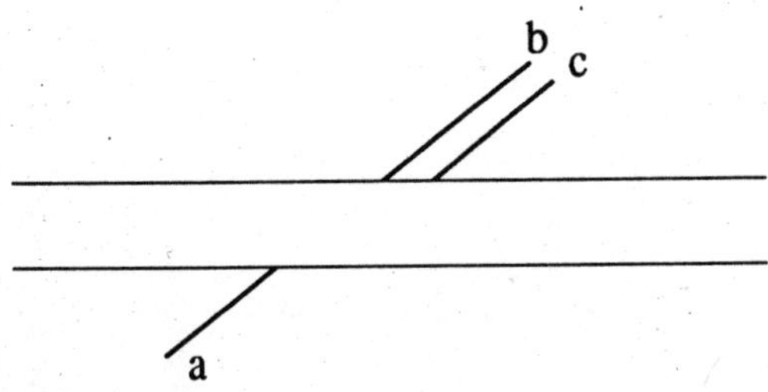


**Two diagonals**

Which of the diagonals is longer: AB or AC?

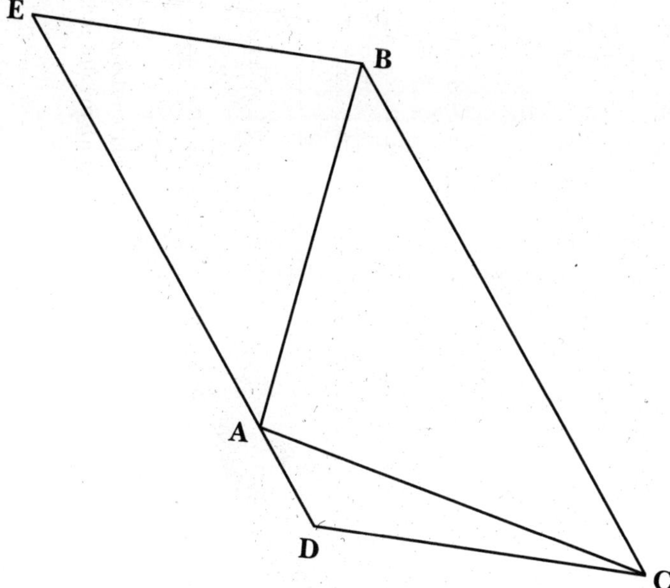

**Muller layer illusion**

Which of the two lines, A or B, is longer?

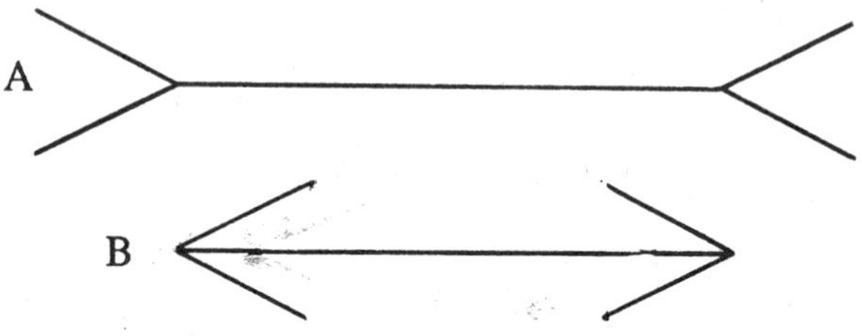


**Larger circles**

Which of the two circles, A or B, is larger?

(Basics of fundamental geometry—Circle inserted in a square and a circle circumscribing a square. Theorem on Circles and applications; Hall and Knights Geometry, p. 378, Blackey edition, London)

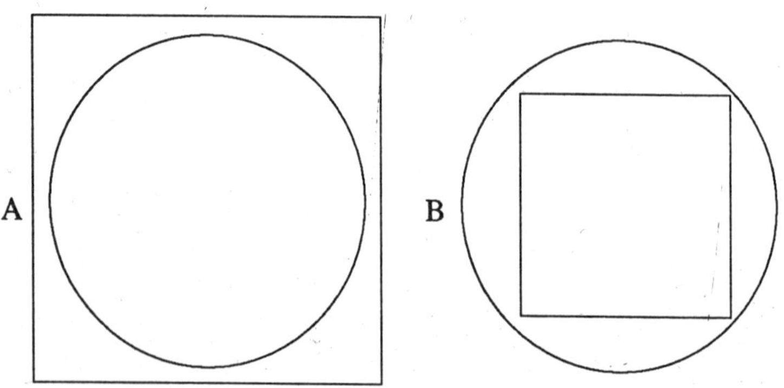

**Distorted circle**

Why does the circle appear distorted? (Circle inscribed in a triangular stub)

**Strange triangle**

Have a look upon this wooden triangle. One can draw it on paper but no carpenter could build it. If he aligns a triangle precisely with squared pieces of wood, its silhouette does simulate this figure, due to illusory effect.

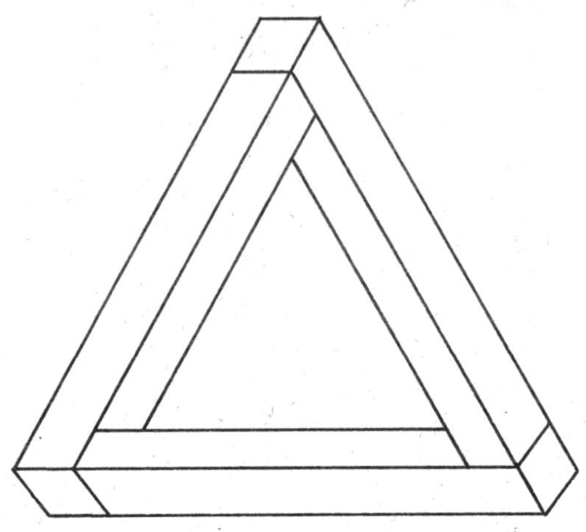

**Lattic structure**

See this frame. Ask any carpenter to dish it out. A master carpenter could hardly do so, despite his skills or hard work.

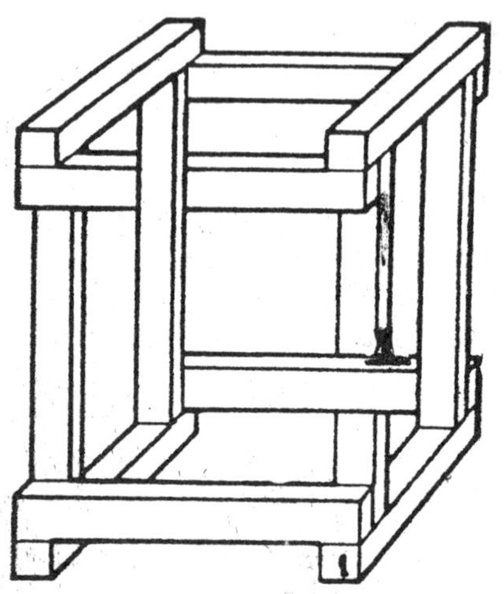

**A missing piece of cake**

In the figure below, a triangular piece of cake is missing. If you turn the figure upside down you will find the missing piece of cake. (Effect of illusion)

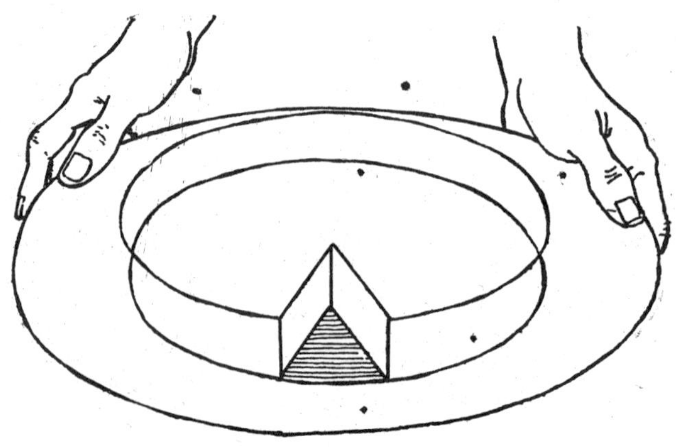

**Positive and negative**

Given in the centre is a photographic negative. Find out which figure is its positive (developed print).

**Disappearing circle**

Bring your nose close to the bottom ring. One of the black circles will disappear (illusion).

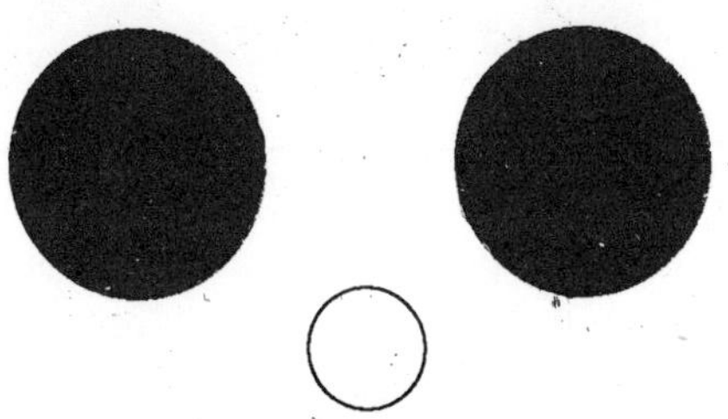

■■

# 11

# AMAZING STORIES ABOUT NUMBERS

1. **Rice story**

Once upon a time, when Japan was ruled by a powerful general, Toyotomi Hideyoshi, there was a so-called clever person specially loved by the general. His name was Sorori Shinzaemon. One day, the general decided to present Sorori a gift for his pithy efforts to win his heart. He asked Sorori, 'what would you like'? Sorori saucily replied "Please give me a grain of rice today, my lord. Then tomorrow, give me two grains and the next day, please give me four more and the day after that eight more and so on for thirty days."

Rice was very scarce in those days because of famine. But Hideyoshi of course granted this apparently upon the entreaty. Well, on the 30th day, Hideyoshi was surprised to see Sorori waiting with a big wrapping cloth, large enough to hold the contents of a whole storage house.

'What are you doing with that,' inquired Hideyoshi. 'Well, my lord!' Sorori explained (with a smile), 'it is for carrying the rice you promised.'

If you do the sum 1+2+4+8+16+......... for thirty days, it adds up to 1073741823 grains of rice. It shall need a whole storage house to hold in. What a surprising story it was about the magic of numbers.

## 2. Distribution of sheep

Once upon a time, there lived a shepherd. He had three sons — Ram, Shyam and Kishan. One day the old man fell ill. He called his three sons and said 'I have 17 sheep for you to take care of. Ram, you take $1/2$, Shyam you take $1/3$ and Kishan you take $1/9$ of the sheep. Share the sheep according to my wish, but you should not kill any one of them for distribution. The old man died after saying this.

The three sons were very said but they were also puzzled as to how they could share the 17 sheep. They could not reach on any conclusion.

A wise man having one sheep of his own passed by and heard of their trouble. He said, 'I can give you my sheep to make 18 in all and that will solve your problem.' Boys were happy.

Now, Ram took $^1/_2$ of the 18 sheep, i.e. 9, Shyam took $^1/_3$ of the 18 sheep, i.e. 6 and Kishan also took $^1/_9$ of the 18 sheep i.e. 2. In this way, the three sons got their treasure, 9+6+2=17 sheep and give the wise man his sheep back.

How strange! Can you discover why the wise man's advice worked. (According to old shepherd, the sum of $^1/_2+^1/_3+^1/_9=^{17}/_{18}$). Therefore the 17 sheep could be distributed among the three sons?

**3. A handful of grains**

Dear readers, you all must have seen a chess board. It is a board which is used to play chess—the National Game of Russia. The board consists of 8 rows and 8 columns. In all, there are 64 squares within (32 black and 32 white).

If you want to ask something from someone, ask for a handful of grains. Give him a chess board and tell him to put one grain on the first square and two on the second.

At each square succeeding, the double is reckoned. Then four grain on third, eight in the fourth, sixteen in the fifth and so on i.e., the grain on each square only double the last.

Do you know the number of grains which one has to put on the last square—The answer is $2^{63}$ and the total number of grains on the board would be $2^{64}+1$ which would be

18, 446, 744, 090, 486, 767, 615 grains.

Eighteen million million million — and someone has calculated that it is equal to the entire wheat production of the whole world for five hundred years.

**4. Height of a pyramid**

Pyramids of Egypt are very famous. About 2500 years ago, an Egyptian king made a declaration if some one can measure the height of the pyramid, he will be given a fabulous award. This work was taken by the Greek mathematician, Thales. He surprised the people by calculating the height of the pyramid from the length of the shadow of a stick.

Thales used the fact that the big triangle ABC formed by the pyramid and its shadow and the small triangle DCE formed by the stick and its shadow, are similar (Fig. above).

Hence

$$\frac{AB}{BC} = \frac{DC}{CE}$$

Thales could have measured the lengths BC, DC and CE and used this equation to calculate the height AB of the pyramid.

5. **Seven bridges**

In the Prussian city of Konigsberg, there were 7 bridges over the river Pregel. For a long time many people tried to take a walk, which involved crossing all the bridges just once without crossing any of them twice. It remained a challenge for a very long time and no one could cross these bridges as per the above condition.

Finally a Swiss mathematician, Leonhard Euler (1707-1783), solved the problem by showing that it is simply impossible to do it. He studied the nature of single line drawings (figures that can be drawn with a single unbroken line without taking the pen from the paper). After that no body tried to cross the bridge.

6. **Forty-nine cows**

An old farmer had 49 cows and seven sons. Each cow was numbered from 1 to 49 by the farmer. The peculiarity of these cows was that each cow gave milk equal to its number i.e. first cow gave one litre of milk, 2nd two litres and so on.

Once the farmer fell sick. He called for his seven sons and asked them to distribute cows among themselves so that each one should gets same number of cows and same amount of milk produced by them. After this the farmer died.

Can you distribute these cows among the seven sons so that each one gets same number of cows and the same amount of milk.

**7. Intelligent shopkeeper**

Most shopkeepers for weighing one kilogram to 40 kg (Integral values only) make use of several weights of 1 kg, 2 kg, 5 kg 10 kg, 20 kg and 40 kg, But in Rampur, there is a very intelligent shopkeeper who makes use of only four weights for weighing 1 kg to 40 kg.

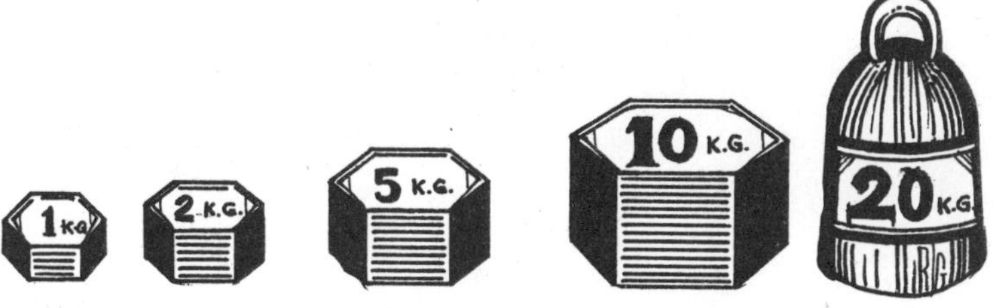


Can you tell what are those four weights which this shopkeeper uses?

# 12

# JUICY PROBLEMS

**Problem 1**

The square of a two-digit number is divided by half the number. After 36 is added to the quotient, this sum is then divided by 2. The digits of the resulting number are the same as those in the original number, but they are in reverse order. The ten's place of the original number is equal to twice the difference between its digits. What is the number?

**Problem 2**

The crescent in the diagram is formed by two ellipses, the larger of which has its centre at C. The width of the crescent between B and D is 9 inches and between E and F is 5 inches (Fig. 12.1). What are the diameters of the two ellipses?

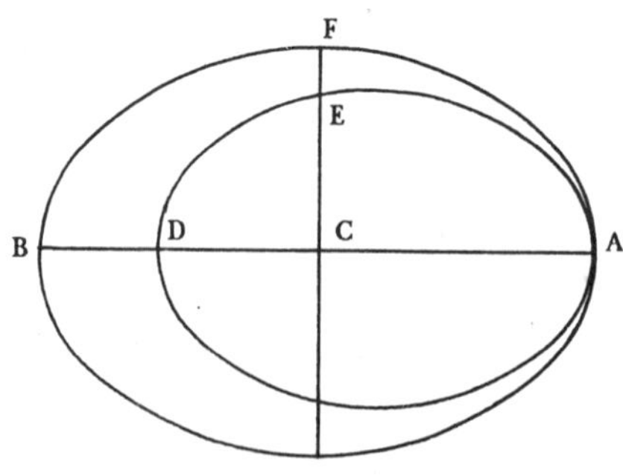


*Fig. 12.1*

*P.S:* Crescent cannot be formed from two circles, since both have same diameter and circumference, unless there is some abnormality behind.

**Problem 3**

Lady Reeta, when asked of her age, replied as 35 years ignoring intervening Saturdays and Sundays in overall span. What was her real age?

**Problem 4**

A man was born in the year 50 BC. How old was he on his birthday in 50 AD?

**Problem 5**

Two fathers and their two sons checked in to a pub to have a drink. The bill brought in was about Rs 3. Each of them spent the same amount and how much did they pay at the counter?

**Problem 6**

A bow and arrow cost Rs 21. The bow cost Rs 20 more than the arrow. What shall be the price of each?

**Problem 7**

In a party, those who handshook arrived at about 28 altogether. How many guests were checked in?

**Problem 8**

Can you tender a rupee note in such a way that there shall be about fifty coins but none of the 2 paise coins?

**Problem 9**

A monkey starts upping a tree 30 ft tall. Each hour, it hops 3 ft and slips back 2 ft. How many days does it take to reach the top (Fig. 12.2).

*Fig. 12.2*

**Problem 10**

At a fruit stall in a market 8 bananas, 7 apples and 3 grape fruits weigh as much as 3 apples, 6 bananas and 6 grape fruits. If a banana weighs two thirds as much as a grape fruit and a dozen apples weigh 3 kg, how much does a grape fruit weigh. (Fig. 12.3)

*Fig. 12.3*

**Problem 11**

One particular year of the last century (1800-1899) if viewed in a mirror, it increased exactly 4.5 times. Which was this year?)

**Problem 12**

The number of my car is such that if you divide it by 2, 3, 4, 5, 6, 7, 8, 9 and 10 you will get a remainder as 1, 2, 3, 4, 5, 6, 7, 8 and 9, respectively. But, if you divide it by 11 then the remainder is zero. Can you tell the number of my car?

**Problem 13**

10 cats ate 10 rats in 10 seconds. How many cats are required to gobble some 100 rats in 100 seconds?

**Problem 14**

Can you imagine a number which is equal to the cube of the sum of its digits?

**Problem 15**

Which number appears same, if read in either the conventional or inverse manner?

**Problem 16**

A boy celebrates his birthday only after every four years. What is his date of birth?

**Problem 17**

A car has a three-digit number which is the square of some number. The other one also has a three-digit square number, but the first digit of the number of first car has become the last digit of the second car. What are the numbers of these two cars?

**Problem 18**

The small hand of a clock is at 12 and the larger one makes an angle of 90° with the smaller one. What is the time in the clock?

**Problem 19**

One cat tells the other that there are two cats in front of me. The other one also cries that he too had two behind him. How many are they?

**Problem 20**

A shopkeeper has six weights and by combining them he can weigh from 1 kg to 364 kg. What are those six weights?

**Problem 21**

Can you find out three digits such that the sum of cubes of each digit gives a three digit number containing the same digits?

**Problem 22**

At 6 O'clock the hands of a clock are exactly opposite each other. When do they oppose again?

**Problem 23**

If a clock in a mirror tells twenty to three, what time it was?

**Problem 24**

What is the area of the shaded portion in Fig. 12.4 when the diameter of each circle is 1 metre.

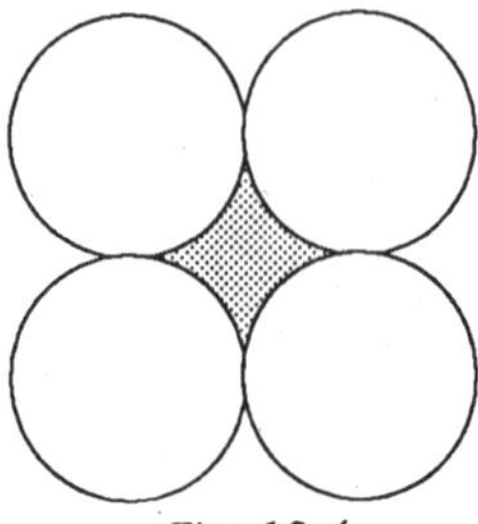

*Fig. 12.4*

**Problem 25**

Which are in maxima — inches in a mile or Sundays in one thousand years?

**Problem 26**

Which are in maxima: seconds in a week or feet in 100 miles?

**Problem 27**

Find a single digit number which when raised to the fifth power is equal to 32768.

**Problem 28**

Which number gives the same result when it is added to 4 as when it is multiplied by 4.

**Problem 29**

What is the sum of all those numbers from 1 to 2000?

**Problem 30**

How many signals could be represented by the three traffic lights, if all possible combinations were allowed, assuming that at least one colour shows?

**Problem 31**

Demonstrate how six sixes (cricket) can equal a gross?

**Problem 32**

A man ate 100 grapes in five days, each day eating 6 more than the previous day. How many grapes did he eat on the first day?

**Problem 33**

How can you represent one by writing nine three times?

**Problem 34**

A carpet has an area of 120 square feet and one diagonal 17 feet. What are the carpet's dimensions?

**Problem 35**

If 8 men can dig 16 holes in 32 days, how long will it take 4 men to dig 8 holes?

**Problem 36**

What number gives you the same result whether you divide or subtract by five?

**Problem 37**

When tomorrow is yesterday, today will be as nearer to Sunday as today was when yesterday was tomorrow. What day was it?

**Problem 38**

What number gives you the same result whether you multiply by 6 or add 6 to it?

**Problem 39**

Find a three-digit number which on reversing the positions of first two digits will yield a number 20 per cent greater to the original?

**Problem 40**

The day before yesterday I was 52. But, next year I will be 55. Can you tell my date of birth and the date on which this statement was made?

# 13

# SOME MORE FUNS

1. A woman went into a shop and bought one shirt and an underwear. The shirt cost Rs 100 more than the underwear. The two together cost Rs 150. How much did each cost?

2. The total weight of a container of flour is 19 kg. After using one third of the flour, the container and flour weigh 14 kg. How much is the weight of the container that holds the flour?

3. How many minutes is it past eight O'clock if 74 minutes ago it was half as many minutes past seven?

4. **To tell the number which your friend has thought:**
   1. Ask your friend to think a number.
   2. Tell him to add seven.
   3. Ask him to multiply by 2 and then subtract 4.

   Ask him to tell you the answer. Now you divide this answer by two and subtract 5. You will get the number which your friend has thought.

   **Example:**
   1. Suppose your friend has thought 20.
   2. On adding 7 it becomes 27.
   3. On multiplying by 2 it becomes 54.
   4. On subtracting 4 it becomes 50.
   5. Now divide 50 by 2 which is equal to 25 and subtracting 5 it becomes 20. This was the number thought by your friend.

5. In fig. 13.1 a beautiful maze is given. You try to enter at the

arrow mark. Can you come out?

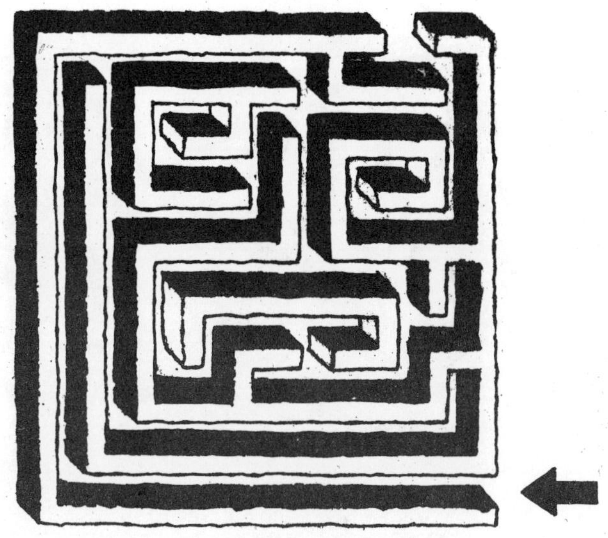

*Fig. 13.1*

6. In fig. 13.2 a circular maze is given. Enter at the arrow mark and come out from the same place.

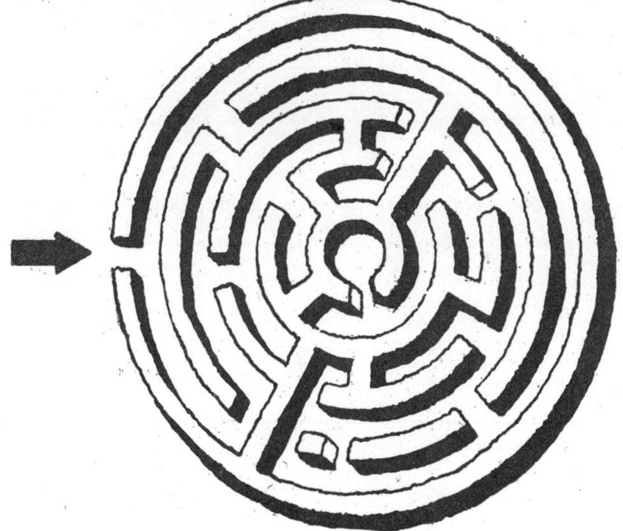

*Fig. 13.2*

7 Figure 13.3 shows a zig-zag maze. Enter at the arrow mark and come out from the same place.

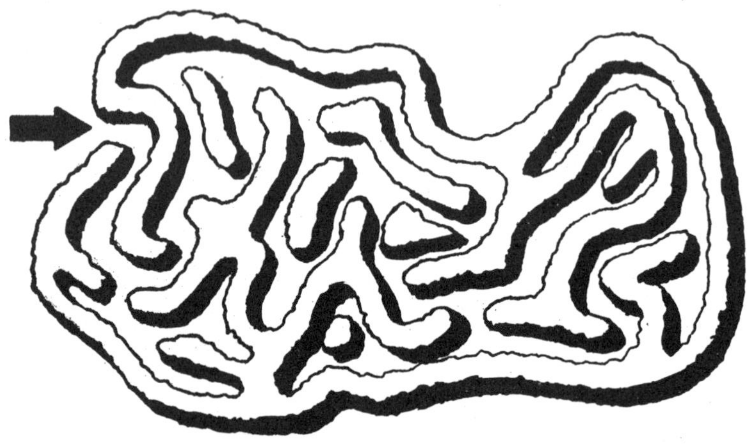

*Fig. 13.3*

8. Complete this magic square.

| | | |
|---|---|---|
| 4 | | 2 |
| | 5 | |
| | | 6 |

9. Complete this magic square such that on adding up numbers down, across or diagnonally the result is always 33.

| | | |
|---|---|---|
| 12 | 7 | 14 |
| | | 9 |
| 8 | | 10 |

10. Given below is a magic square of 5×5. Complete it.

| | | | | |
|---|---|---|---|---|
| 15 | 8 | 1 | 24 | |
| 16 | | 7 | | 23 |
| | 20 | | 6 | 4 |
| 3 | 21 | 19 | 12 | |
| 9 | 2 | 25 | 18 | 11 |

11. Given below (Fig. 13.4) are 15 matches. These have been arranged five squares. Remove three of the matches and turn the five squares into three.

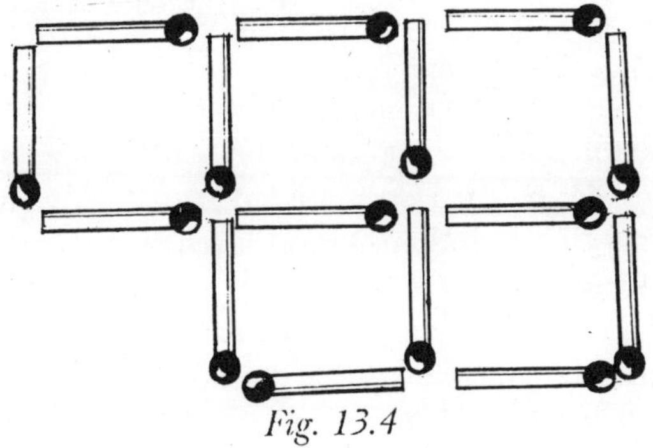

*Fig. 13.4*

12. Put six coins of the same type on to a table in the pattern shown in fig. 13.5. Can you, by moving only three of the coins one at a time, turn the two lines into a circle?

*Fig. 13.5*

13. **How to tell the year marked on a coin?**
Ask your friend to take a coin out of his pocket and to tell you the third figure in the date of the coin.

Now double it.
Add 5.
Multiply by 5.
Add the last figure to your result.
Subtract 25 from the total.
You will have the year on the coin.
**Example:**
Suppose a coin has a marking of 1985.
The third figure is 8.
Its double is 16.
On adding 5, it becomes 21.
On multiplying by 5 we get 105.
Adding the last figure we get 105+5=110.
Now subtracting 25, we get 110–25=85, so the year marked on the coin is 1985.

14. **Can you tell me the answers of the following problem?**
Ram and Shyam learn German and French. Shyam and Kamal learn French and Italian. Vinod learns Russian and German and Ram also takes Russian.
(a) Who learns German but not French?
(b) Which language does Shyam not learn?
(c) Who learns German, French and Italian?
(d) Who learns Russian but not French?

15. You have 8 matches as shown in fig. 13.6. How would you write six with them?

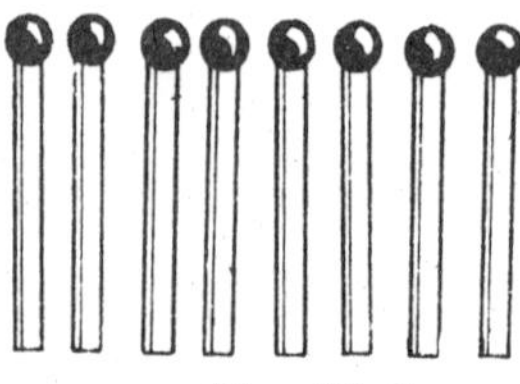

*Fig. 13.6*

16. You have 10 matches, as shown in fig. 13.7. How can arrange them to make FIVE?

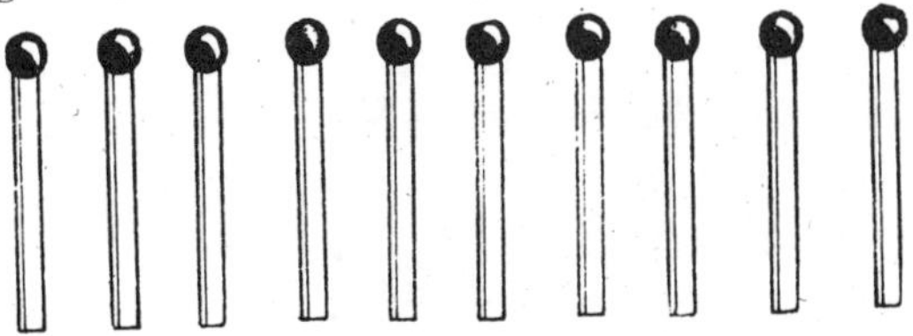

*Fig. 13.7*

17. Given below are two figures (fig. 13.8). Can you work out which one has more number of squares?

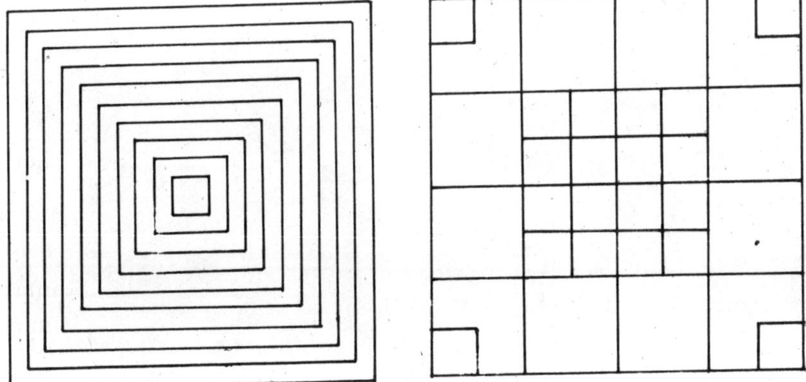

*Fig. 13.8*

18. **The answer is always ten.**
    1. Ask your friend to think of a number.
    2. Add 9 to the number.
    3. Double your answer.
    4. Add 3 to it.
    5. Now multiply it by 3.
    6. Subtract 3 from the answer.
    7. Divide it by 6.
    8. Subtract the number you thought.
    9. The answer is ten.

**Example:**

1. Suppose the number thought by your friend is 8.
2. On adding 9 to it, we get 17.
3. Double of 17 is 34.
4. Adding 3 to it, we get 37.
5. Multiplying by three, it becomes 111.
6. Subtracting 3, we get 108.
7. Divide by 6, we get 18.
8. Subtracting the original number 8, we get 10.

So the answer is always 10.

19. Here are three triangles (Fig. 13.9). By moving only three matches, make five triangles.

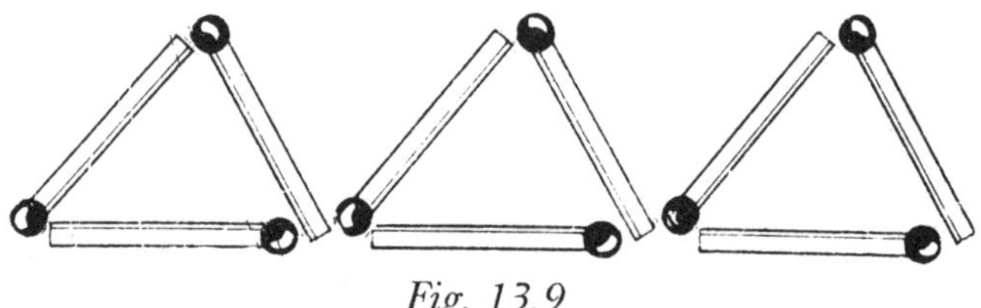

*Fig. 13.9*

20. When I am as old as my brother is now, my sister will be three years older than I am at the moment, and my brother will be 12 years older than my sister at the moment. If our ages add up to 57 now, how old are we (Fig. 13.10)?

*Fig. 13.10*

21. My father is 21 years older than me. In 12 years time his age will be twice mine. How old am I now?

22. If my sister is 21 and she is 8 years older than me and my brother is 15 and he is two years older than me, how old am I?

23. Two mothers and two daughters went into a sweet shop and spent six rupees. If each of them spent the same amount, how much did they each spend?

24. In four years time Sita will be twice as old as she was 11 years ago. How old is she?

25. Can you make 20 using only two 3s?

26. Can you make 100 using only four 6s?

27. Can you make 1000 using only eight 8'?

28. Find out the correct digit for each letter:

| A | B | C |
|---|---|---|
| –4 | 5 | 9 |
| C | A | B |

29. Place 10 coins as shown in the figure 13.11 below. By moving three coins make the pyramid upside down.

*Fig. 13.11*

30. Arrange 16 coins as shown below in fig. 13.12.

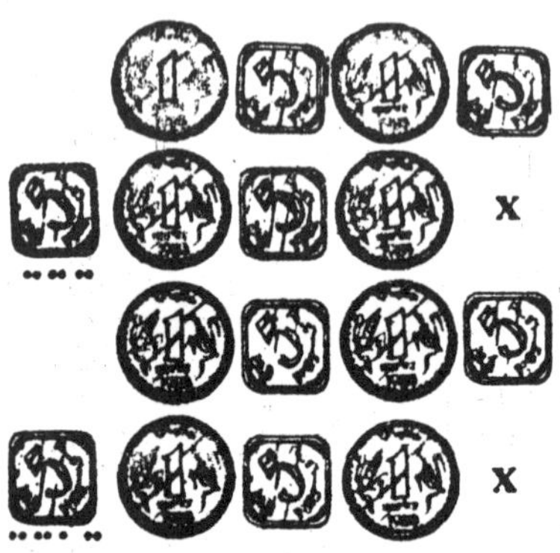

*Fig. 13.12*

Now arrange in such a way that they are alternately placed vertically and horizontally. You can touch only two of the coins.

31. If it takes four minutes to boil one egg, how long will it take to boil three eggs?

32. Given below are 16 dots (fig. 13.13). Join them with six straight lines without lifting your pencil from the paper.

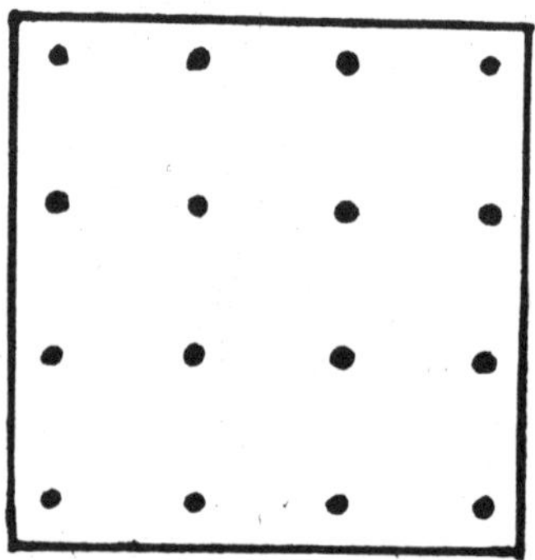

*Fig. 13.13*

33. What two numbers multiplied together will be equal to 7?

34. Convert a circle of area 100 square meters into a square of equal area. What would be the side of square?

35. Divide the triangle given below in fig. 13.14 into three parts and make two perfect squares.

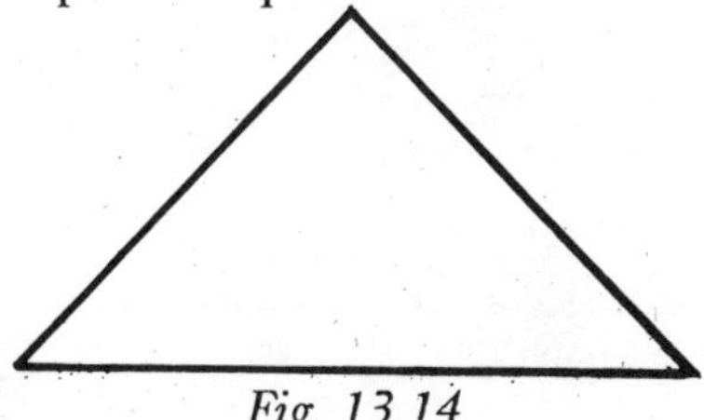

*Fig. 13.14*

36. Draw 8 straight lines to connect all the dots of fig. 13.15 so that three stars shown below can be placed in three squares.

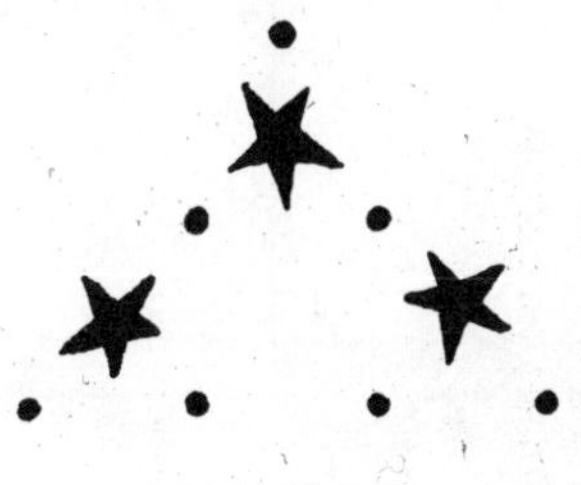

*Fig. 13.15*

37. What number which when increased by 3 is exactly divisible by 12, 15, 18 and 24?

38. The difference between the number and its square root is 90. What is the number?

39. Thirty divided by half and minus thirty. What is the result?

40. If you divide 60 by half and then add 40, what will be the result?

41. **Find the value of X.**

| | | | | |
|---|---|---|---|---|
| 2 | 1 | 8 | 5 | 9 |
| 3 | 7 | 2 | 6 | 2 |
| 4 | 2 | 1 | 1 | X |

42. **Mention the word represented by X in the following formula.**

31 31 X 31

43. **Find the value of X and Y.**

7 8 6 9 5 10 X Y 3 12

44. Write 37 on a peice of paper and keep the same in your pocket. Let someone suggest any number formed of three identical figures and add up the three figures and divide the orignal number by the answer.

45. **What are X and Y?**

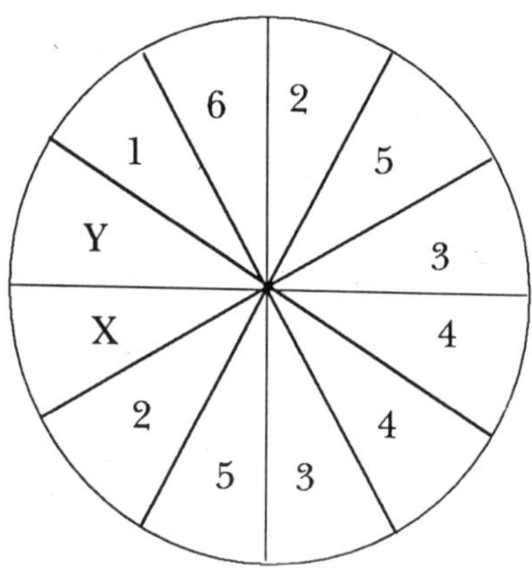


■■

# ANSWERS

## CHAPTER 5

**Fun 1** — 36

**Fun 2** — 38

**Fun 3** — 76

**Fun 4** — 27

**Fun 5** — 14

**Fun 6** — 11

**Fun 7** — 5

**Fun 8** — 13

**Fun 9** — 47

**Fun 10** — 20

**Fun 11** — 35

**Fun 12** — 48

**Fun 13** — The shaded part of the top shows 6 cubes, while that of the bottom shows 7 cubes.

**Fun 14** — 25 (twenty five) mangoes can be arranged in 12 rows in three ways as shown in Figures 5.14, 5.15 and 5.16

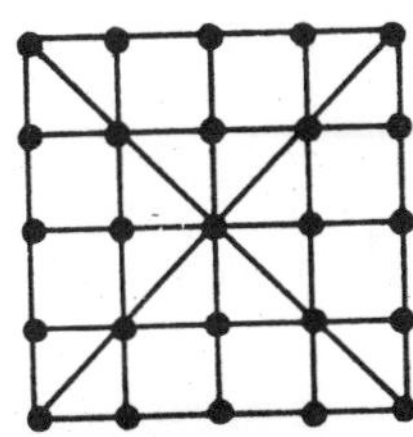

*Fig. 5.14*

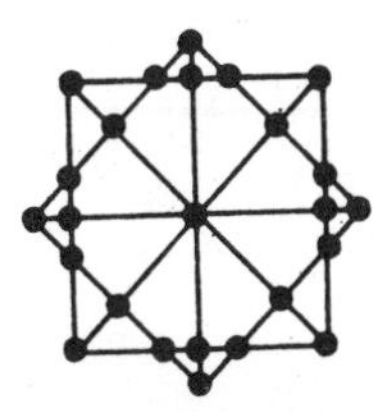

*Fig. 5.15*

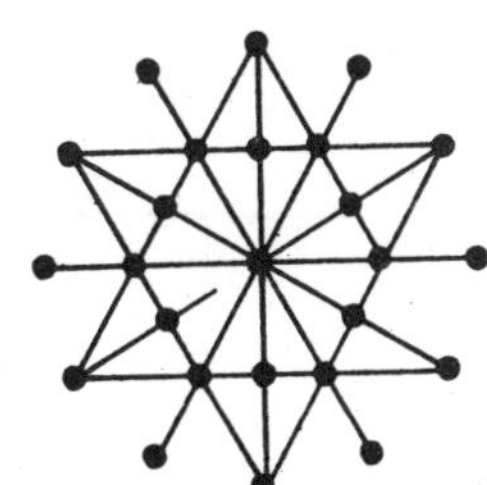

*Fig. 5.16*

## CHAPTER 6

**Fun 1** — 6

**Fun 2** — 20

**Fun 3** — 5

**Fun 4** — 15

**Fun 5** — 15

**Fun 6** —

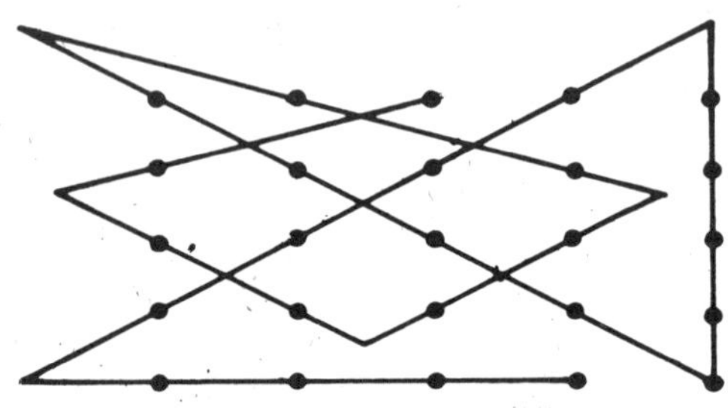

**Fun 7** —

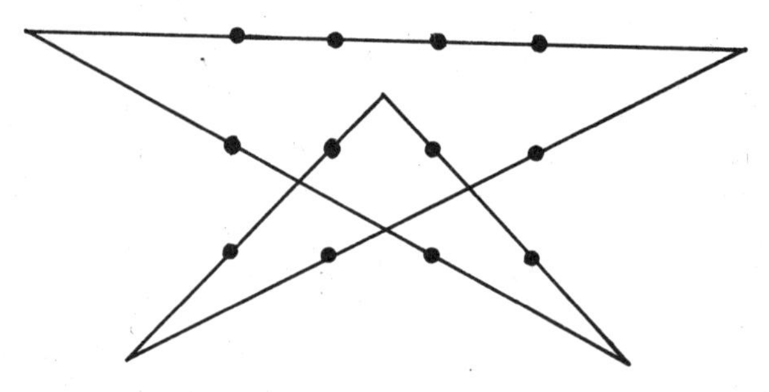

## CHAPTER 7

**Fun 1** —

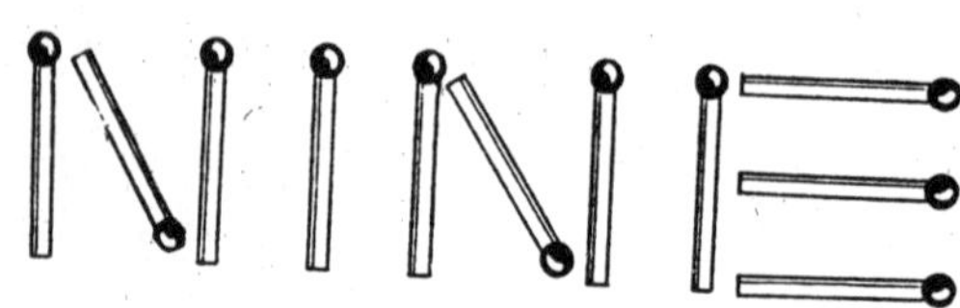

**Fun 2** —

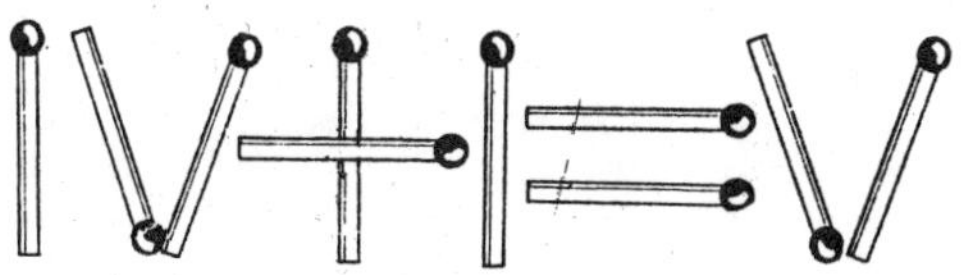

**Fun 3** —

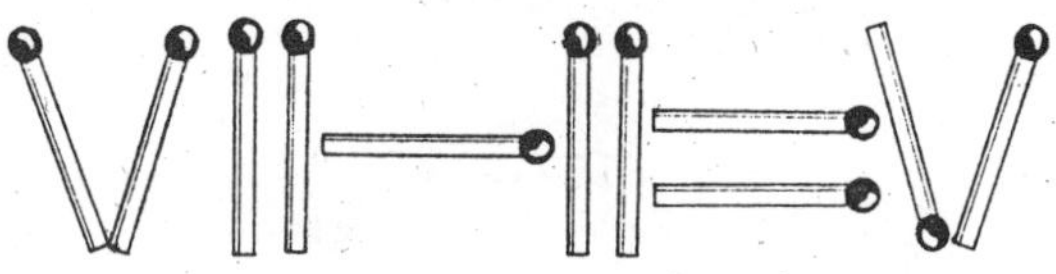

**Fun 4** —

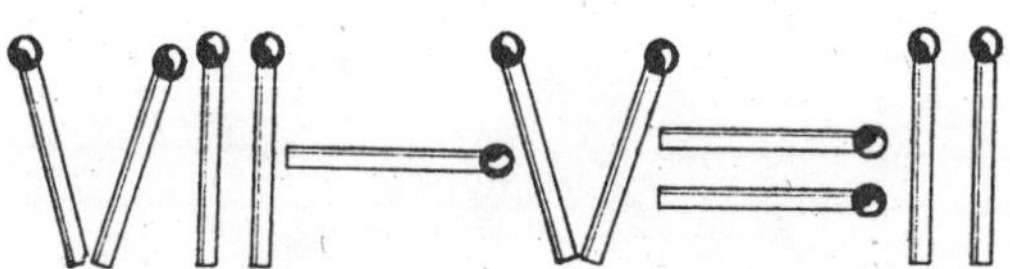

**Fun 5** —

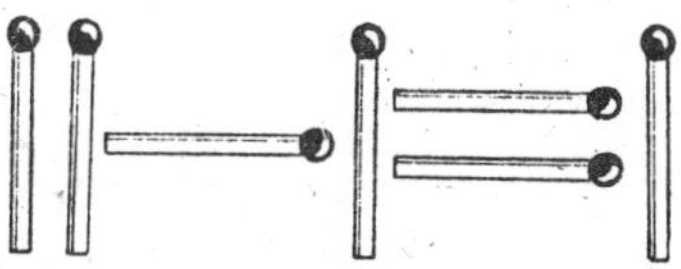

**Fun 6** — Turn the first digit to make it 0.

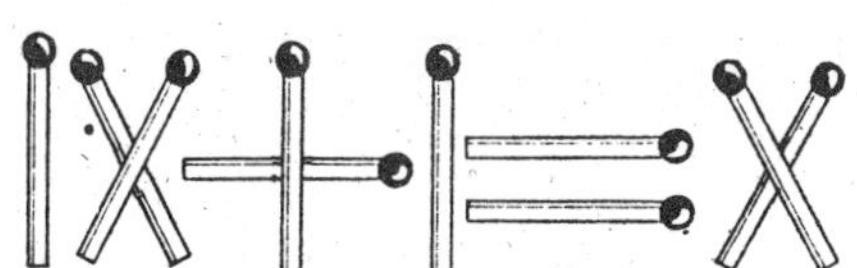

**Fun 7** —

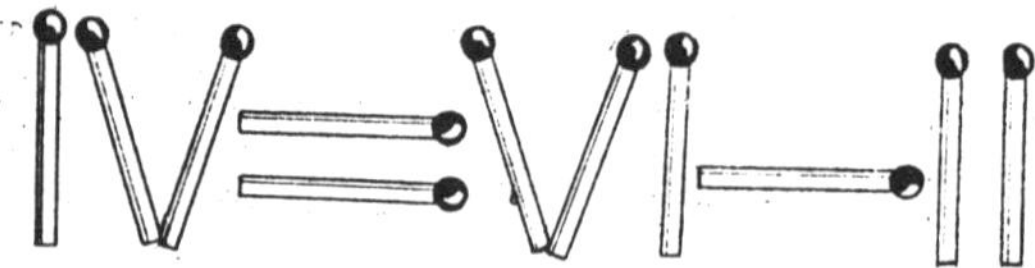

**Fun 8** —

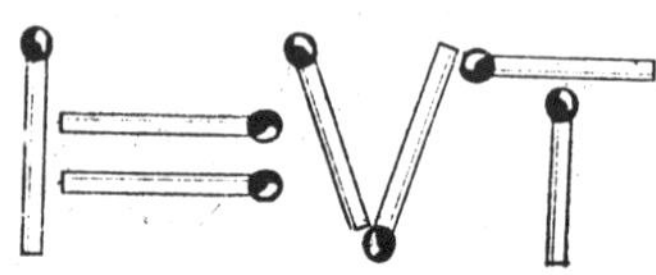

**Fun 9** —

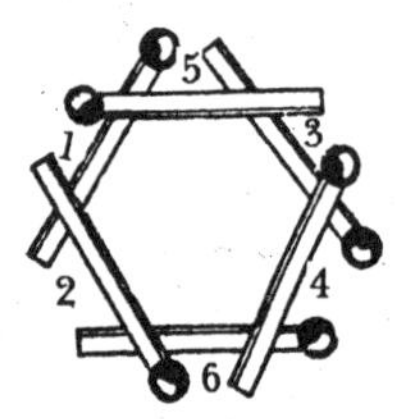


**Fun 10** —

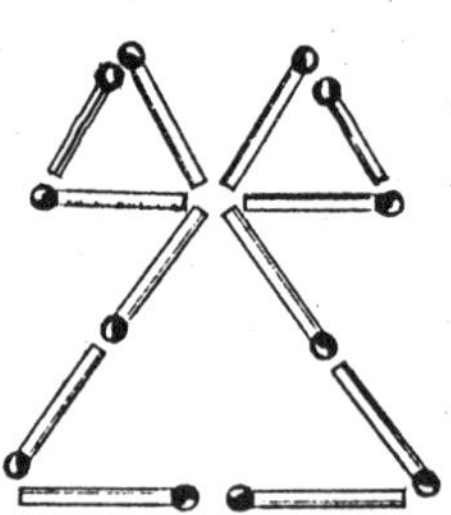

**Fun 11** — Eight triangles: 8–8–1; 8–7–2; 8–6–3; 8–5–4; 7–7–3; 7–6–4; 7–5–5; 6–6–5.

**Fun 12** —

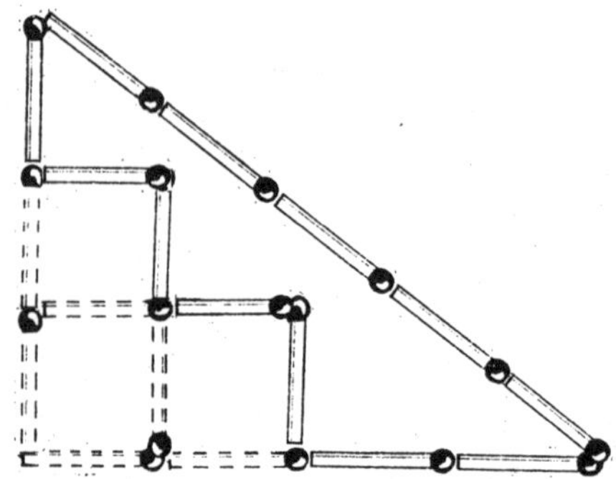

# CHAPTER 8

**Magic Square 1**

| | | | | |
|---|---|---|---|---|
| 25 | 32 | 9 | 16 | 23 |
| 31 | 13 | 15 | 22 | 24 |
| 12 | 14 | 21 | 28 | 30 |
| 18 | 20 | 27 | 29 | 11 |
| 19 | 26 | 33 | 10 | 17 |

**Magic Square 2**

| | | | |
|---|---|---|---|
| 45 | 32 | 31 | 42 |
| 34 | 39 | 40 | 37 |
| 38 | 35 | 36 | 41 |
| 33 | 44 | 43 | 30 |

**Magic Square 3**

| | | | |
|---|---|---|---|
| 20 | 7 | 6 | 17 |
| 9 | 14 | 15 | 12 |
| 13 | 10 | 11 | 16 |
| 8 | 19 | 18 | 5 |

**Magic Square 4**

| | | | |
|---|---|---|---|
| 1 | 14 | 15 | 4 |
| 8 | 11 | 10 | 5 |
| 12 | 7 | 6 | 9 |
| 13 | 2 | 3 | 16 |

**Magic Square 5**

| | | | | |
|---|---|---|---|---|
| 37 | 44 | 21 | 28 | 35 |
| 43 | 25 | 27 | 34 | 36 |
| 24 | 26 | 33 | 40 | 42 |
| 30 | 32 | 39 | 41 | 23 |
| 31 | 38 | 45 | 22 | 29 |

**Nasik Magic Square**

| | | | | |
|---|---|---|---|---|
| 9 | 11 | 18 | 5 | 22 |
| 3 | 25 | 7 | 14 | 16 |
| 12 | 19 | 1 | 23 | 10 |
| 21 | 8 | 15 | 17 | 4 |
| 20 | 2 | 24 | 6 | 13 |

# CHAPTER 9

## 1. Magic Star

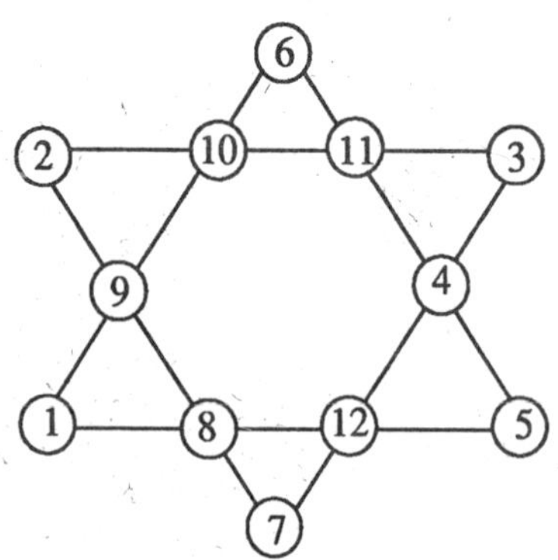

## 2. Honey Comb

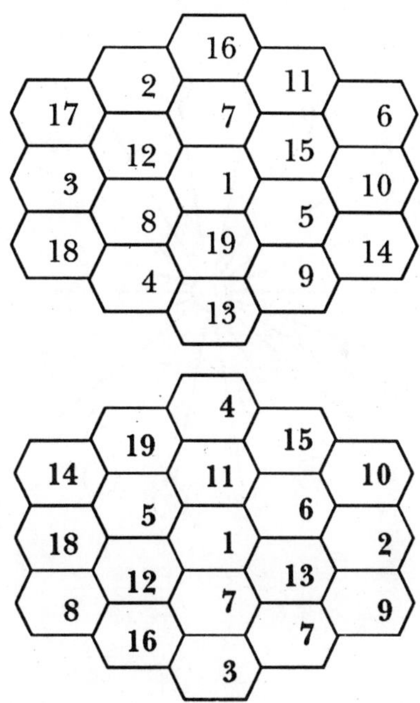


## 3. Magic Figure

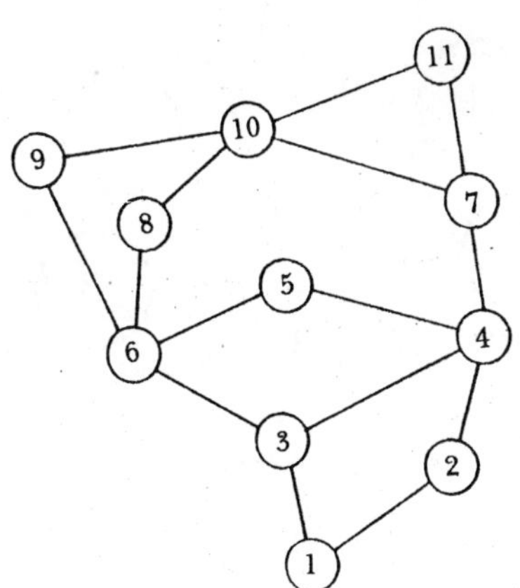

### 4. Adding Wheel

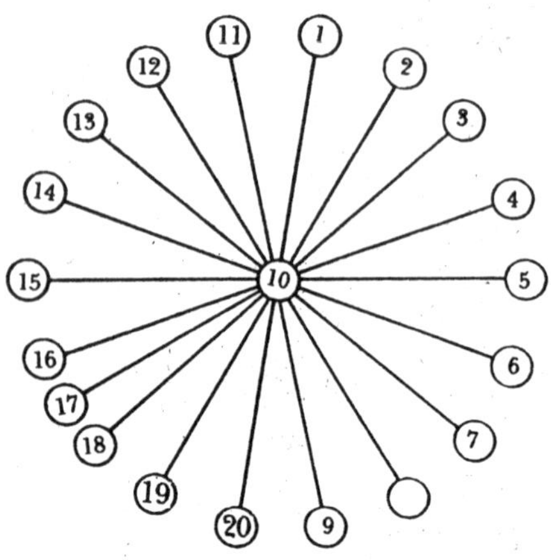


### 5. Magic Circles

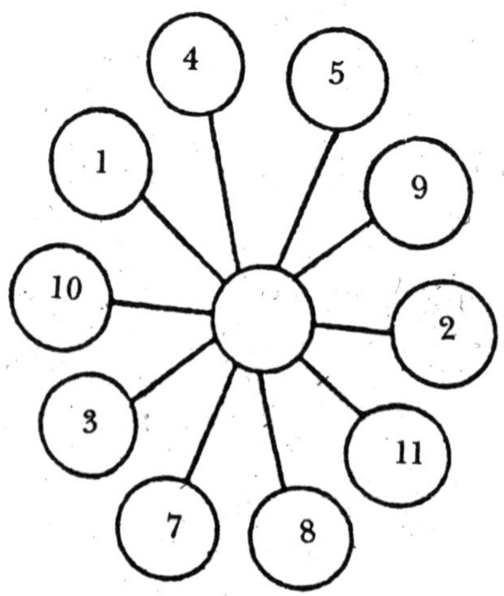

## 6. Number Wheel

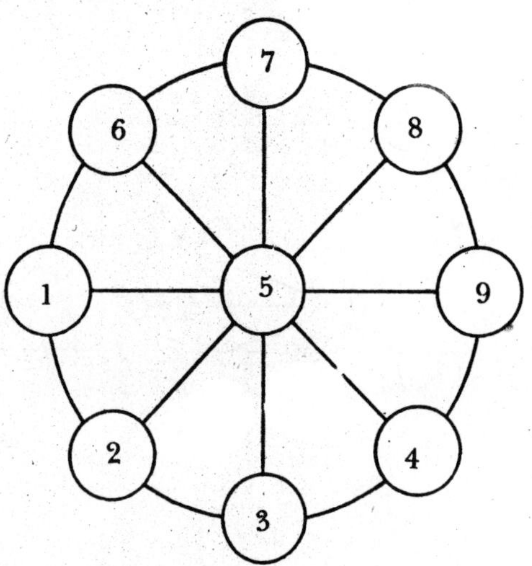


## 7. Eight-Point Star

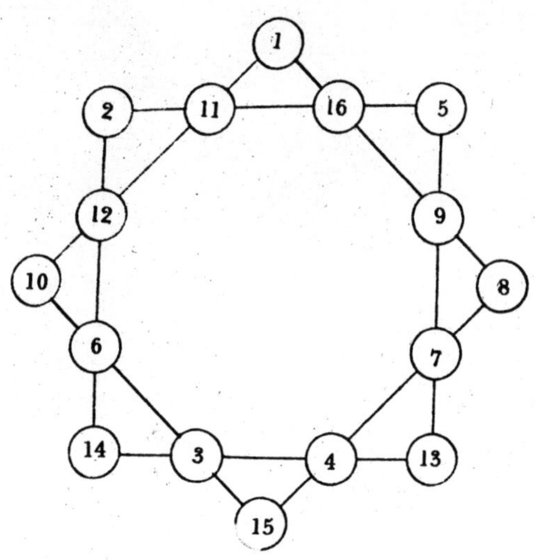

## 8. The Magic Hexagon

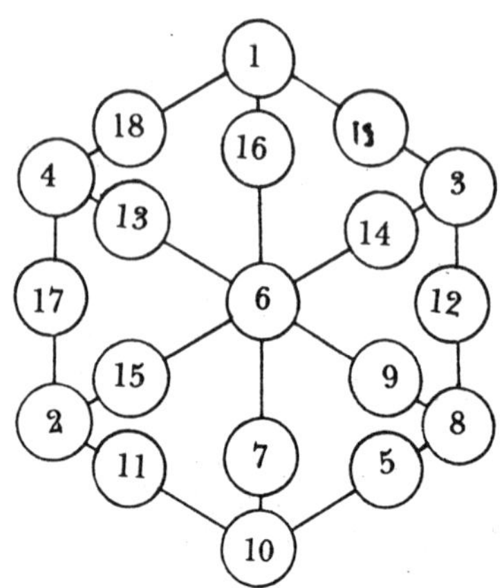


## 9. Missing Number

| | | |
|---|---|---|
| 8 | 15 | 10 |
| 13 | 11 | 9 |
| 12 | 7 | ? |

## 10. Penta Star

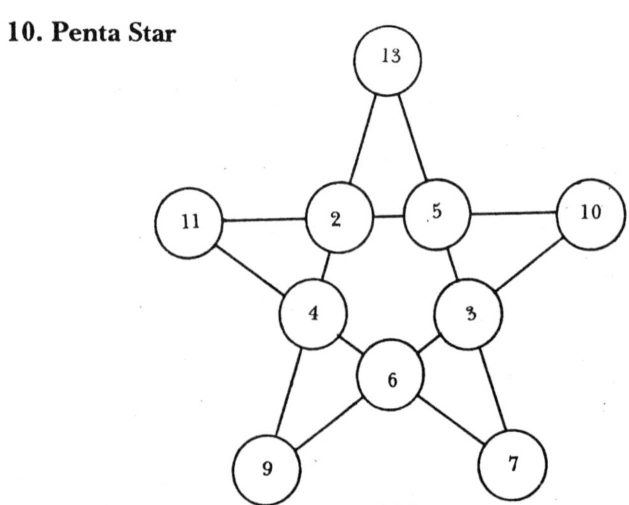

**11. Magic Circle**

Missing ones are 3 and 89. It follows a sequence 0+1=1, 1+1=2, 2+3=5 and so on. Answer is in geometric series.

**12. Magic Movement**

Turn the box containing 9 in block B upside down to make Sum of each block is 18.

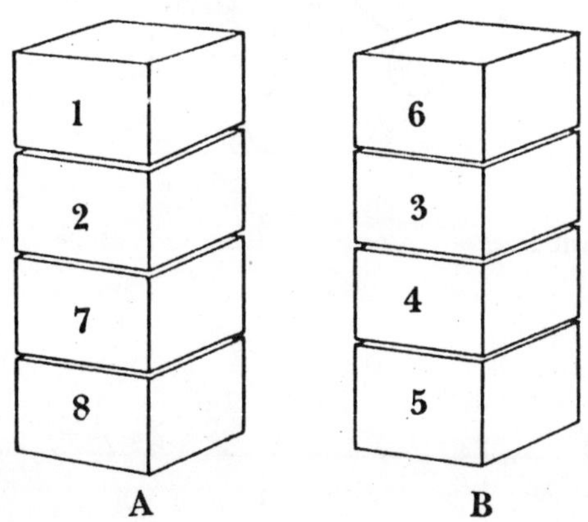


**13. Strange Star**

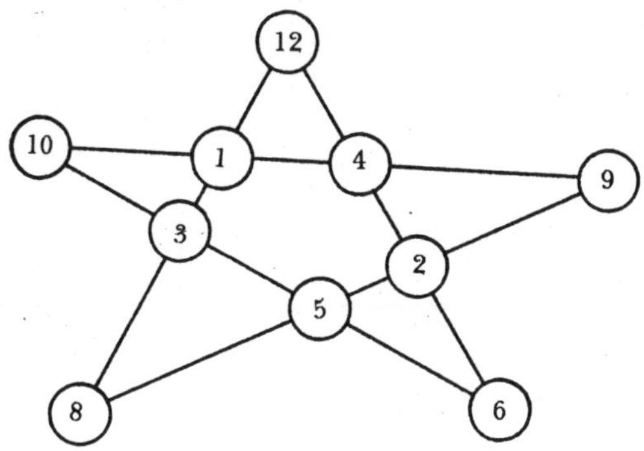

## 14. Lollipop Game

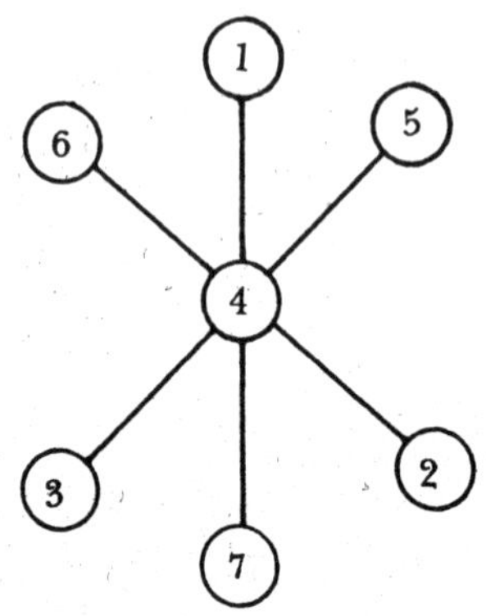


## 15. Seven Point Star

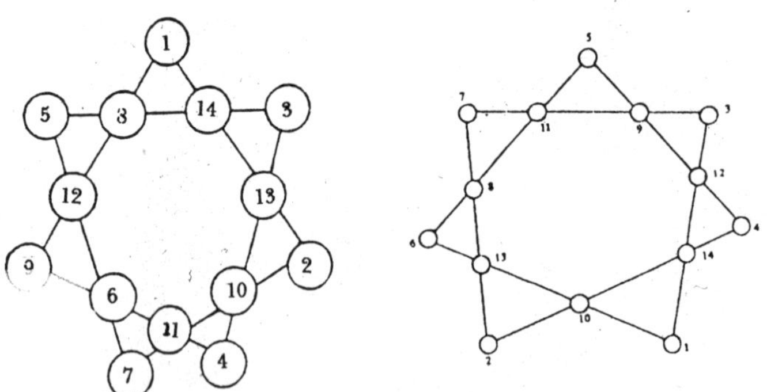

## 16. Eight Point Star

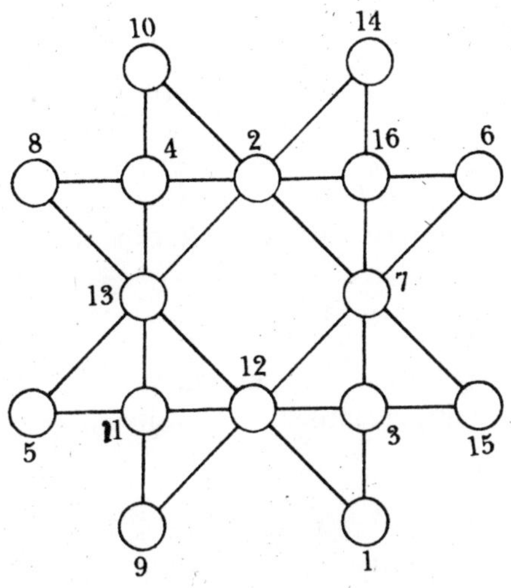


# CHAPTER 10

**Longer Line**

A appears longer than B due to angular effect but both have straight lines of similar lengths.

**Dark and Bright**

White appears larger than the black circle due to irradiation effect but white sphere is etched out of the black steroid.

**Two Perpendicular Lines**

Both the lines bear the same length but the vertical one appears longer to the horizontal.

**Longest Line**

All the three lines, a , b and c are of the same length. Due to different angles, the straight parts of them appear to be of unequal lengths. (See Cantilever principle to justify — the amount of bending during heating creates elongational strain and is proportional to the amount of applied stress. But, it remains the same after cooling, however the marginal value of elongation does play an important role in the Theory of Thermodynamics).

**Intersecting Lines**

Line 'a' appears aligned with 'b' but in fact 'a' is aligned with 'c'. It is a common phenomenon when an iron/wooden rod is dipped in a pool of water, illusion causes a bend in the part of the rod which is under water. Refractive indices of water and atmosphere are different and that difference makes the observer to see a rod bent.

**Two Diagonals**

AB and AC both are of the equal lengths but due to different angles AB appears longer than AC. You can check with a foot-ruler, if you do not believe it.

**Muller Layer Illusion**

A an B are of similar lengths but the one with the outward turning ends appears to be appreciably longer following its obtuse formation.

**Larger Circles**

A looks a little larger in magnitude when compared to B but both circles are of the same size. It is their position that deceives us. Remember the analogy, a square peg cannot fit in a round whole and vice-versa, but the Nature has its proof, by virtue.

**Distorted Circle**

The circle is perfect, but appears distorted because of the background effect of regular triangles, overlapping one another.

**Positive and Negative**

d

## CHAPTER 11

6. **Forty-nine Cows**

Total number of cows $= 49$

Each person gets $=$ 7 cows

Total milk produced $= \dfrac{n(n+1)}{2}$

$= \dfrac{49 \times 50}{2}$

$=$ 1225 litres

i.e. each person should get seven cows producing 175 litres of milk, so the distribution follows:

| | | | | | | | |
|---|---|---|---|---|---|---|---|
| First son | 1, | 11, | 20, | 22, | 24, | 48, | 49 |
| Second son | 2, | 10, | 14, | 23, | 33, | 46, | 47 |
| Third son | 3, | 7, | 12, | 31, | 32, | 35, | 45 |
| Fourth son | 4, | 8, | 17, | 26, | 36, | 41, | 43 |
| Fifth son | 5, | 16, | 19, | 29, | 30, | 34, | 42 |
| Sixth son | 6, | 15, | 18, | 27, | 28, | 37, | 44 |
| Seventh son | 9, | 13, | 21, | 25, | 38, | 39, | 40 |

**7. Intelligent Shopkeeper**
1 kg, 3 kg, 9 kg, 27 kg.

# CHAPTER 12

**Problem 1** — 46

**Problem 2** — Suppose the dia of the larger ellipse is x. So CD=x–9 and EC=x–5. Then x–5 is a mean proportion between x–9 and x, from which x must be equal to 25. Therefore, the diameters of the two ellipses are 50 and 41 inches.

**Problem 3** — 49 years

**Problem 4** — 99 years old. There was no year 0.

**Problem 5** — Each paid Re 1. Obviously, there was a grandfather, father and son.

**Problem 6** — The bow cost Rs 20.50 with the arrow a meagre 50 paise.

**Problem 7** — 8 guests

**Problem 8** — 45 one paise, 2 five paise, 2 ten paise and 1 twentyfive paise coins.

**Problem 9** — 28 hours

**Problem 10** — One grape fruit weighs 600 gms.

**Problem 11** — 1818

6.

18 18

81 81

**Problem 12** — 2519

**Problem 13** — 10

**Problem 14** — 4913 is the cube of 17 which is the sum of its digits.

**Problem 15** — 1961

**Problem 16** — 29 February

**Problem 17** — 196 and 961

**Problem 18** — Fifteen minutes past twelve.

**Problem 19** — Three

**Problem 20** — 1 kg, 3 kg, 9 kg, 27 kg, 81 kg and 243 kg.

**Problem 21** — $1^3+5^3+3^3=153$

**Problem 22** — At $5\frac{1}{2}$ minutes past seven.

**Problem 23** — Twenty past nine

**Probelm 24** = $1-\frac{\pi}{4}$ sq. m

**Problem 25** — Inches in a mile, 63,360 to about 52,000

**Problem 26** — Seconds in a week, 604,800 to 528,000

**Problem 27** — Eight

**Problem 28** — $1\frac{1}{3}$

**Problem 29** — 2,001,000

**Problem 30** — Seven — red by itself, green by itself, amber by itself, red and green, red and amber, amber and green, red and green and amber.

**Problem 31** — 66+66+6+6=144

**Problem 32** — Eight.

**Problem 33** — $9^{9-9} = 9^0=1$

**Problem 34** — 8 feet and 15 feet.

**Problem 35** — 32 days

**Problem 36** — 6.25

**Problem 37** — Sunday

**Problem 38** — 1.2

**Problem 39** — 450.

**Problem 40** — December 31 and statement was made on January First

# CHAPTER 13

1. The shirt cost Rs. 125/- and underwear Rs. 25/-
2. This question can be solved by using simultaneous equations.

   The flour container weighs 4 kg.
3. It is 28 minutes past eight.

8.

| | | |
|---|---|---|
| 4 | 9 | 2 |
| 3 | 5 | 7 |
| 8 | 1 | 6 |

9.

| | | |
|---|---|---|
| 12 | 7 | 14 |
| 13 | 11 | 9 |
| 8 | 15 | 10 |

10.

| | | | | |
|---|---|---|---|---|
| 15 | 8 | 1 | 24 | 17 |
| 16 | 14 | 7 | 5 | 23 |
| 22 | 20 | 13 | 6 | 4 |
| 3 | 21 | 19 | 12 | 10 |
| 9 | 2 | 25 | 18 | 11 |

11.

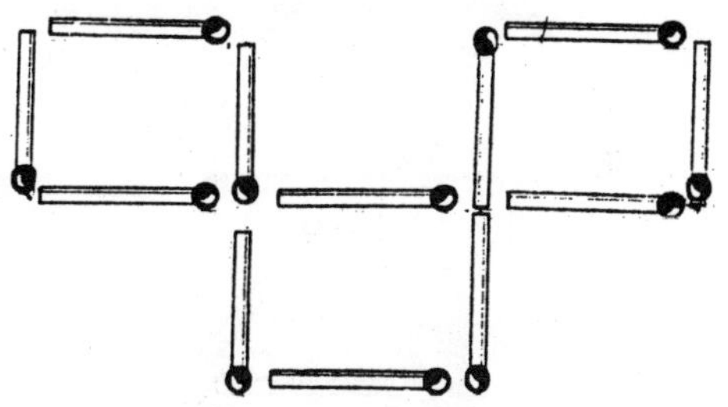

12. Move 4 to touch 5 and 6
Move to touch 1 and 2
Move 1 to touch 4 and 5
(see figure)

14. (a) Vinod (b) Russian (c) Shyam (d) Vinod

15.

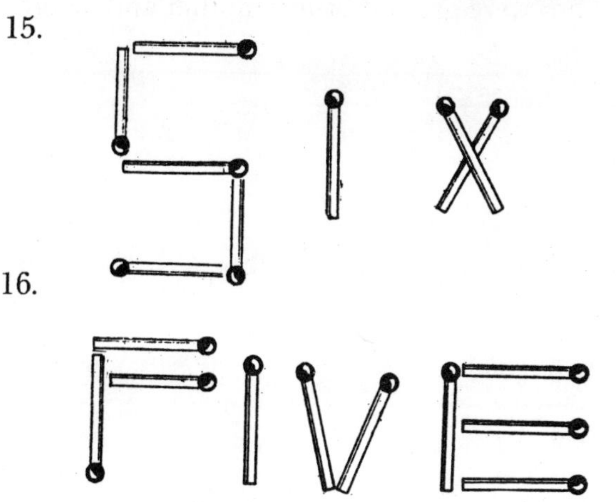

16.

17. Second figure.

19.

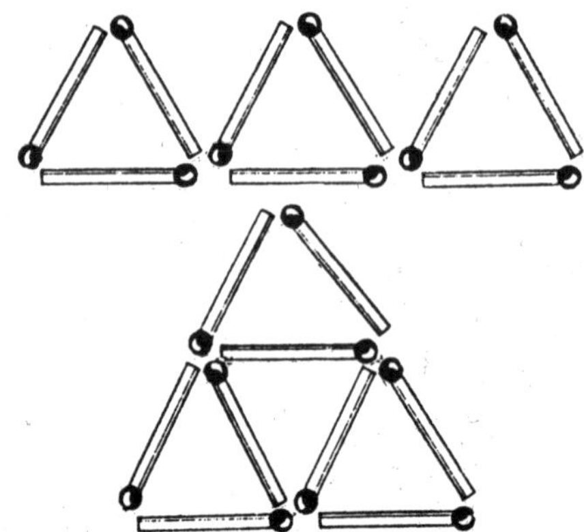

20. Use simultaneous equations to solve this problem. I am 18, my brother is 23 and my sister is 16.

21. I am 9.

22. I am 13.

23. Rs 2 each. They were grandmother, mother and daughter.

24. Sita is 26.

25. $\frac{\lfloor 3}{\cdot 3} = 20$

26. $\frac{66}{\cdot 66} = 100$

27. 888+88+8+8+8=1000

28. A=9, B=5, C=4

29. Suppose we have 10 coins as

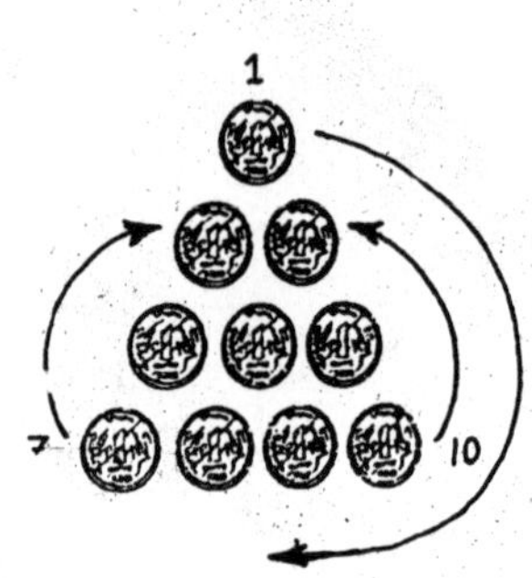

Then move coin no. 7 to the left of coin no. 2. Coin no. 10 to the right of coin no. 3. Then move coin no. 1 to below the bottom row under coin no. 8 and 9.

30. Take x mark coins at a time and move them to the dotted position and push the other coins in these two rows to the vacant places.

31. 4 minutes

32.

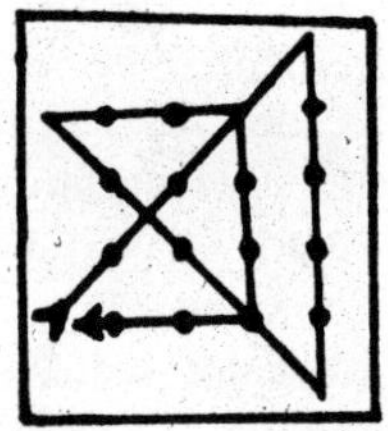

34. 10 metres

35.

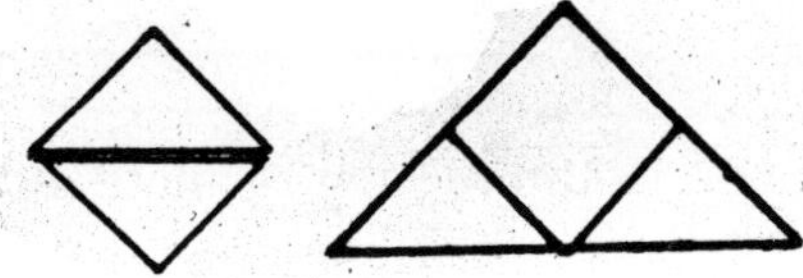

36.

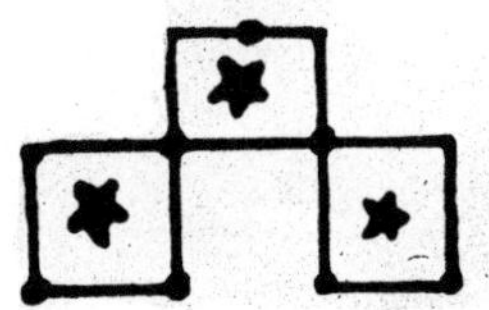

37. 357.

LCM of 12, 15, 18 and 24 is 360.

Subtract 3 from 360 i.e. $360-3=357$.

38. 100.

39. 30.

40. 160.

41. The first column totals 9, second 10, third 11; fourth 12 and accordingly fifth should be 13. Hence the value of X is 2.

42. September. The figures represent the number of days in months. September is the only month which has two 31-day months before and one after it.

43. Make two alternate series starting from 7 6 5 4 3 and 8 9 10 11 12. Accordinly the value of X = 4 and Y = 11.

44. Whatever number he chose you will always have the answer 37.

**Example:**

(1) 444 divided by 4+4+4=37

(2) 666 divided by 6+6+6=37

(3) 999 divided by 9+9+9=37

45. X = 6, Y= 1.

Starting from no. 1 and moving to alternate segments clockwise:

1 2 3 4 5 6

Starting at no. 6 and moving in the same way:

6 5 4 3 2 1.

■■